The Naked Bonobo

Lynn Saxon

Copyright © 2016 Lynn Saxon

ISBN-10: 1523945516
ISBN-13: 978-1523945511

Contents

I wanna be like you-oo-oo

When the chimpanzee fell from grace as the peaceful vegetarian, who would have thought that lurking in the shadows was another hairy cousin, and a much better one at that. This hippie starlet, quietly waiting in the wings, was so much like the chimpanzee that for some time she was known (to those few who knew her at all) as the pygmy chimpanzee. After decades of obscurity, this budding star gradually (over even more decades) made her presence known. Now, finally, her breakthrough has come, and her cue to take centre stage has arrived. So, if you don't want to be a chimpanzee – that nasty, brutish ape – you can now choose the new and improved version: the peaceful and free-lovin' bonobo. Cue Lights, Camera, and...Action!

From virtually all social media talk, it seems that what everyone absolutely knows about the bonobo is that it is all about the sex. Whether it's an image of one great big cuddly love-in or, more bluntly, everyone simply banging everyone else, it is the sex that has led the bonobo down the path to a peaceful, easy-going bliss. Watching bonobos at feeding time at the zoo, or even the videos of them at artificial feeding sites in the wild, it is hard not to be bemused by all their sexual antics. This is how they avoid conflict, we are told; surely this is how we too can do the same. Those chimpanzee cousins of ours don't do this kind of thing – and we all know how horrible they can be.

So today there is the most enticing, lascivious light shining through the dark cloud that hangs over human nature. This is the mother of all indecent exposures: our inner bonobo. If only we can set her free we will wipe out war and conflict and violence. Can any of us refuse the call to release this joyous sexual panacea?

If only it were that simple...

The Bonobo.
Sexy. Gentle. Friendly.
The "make love not war" ape. At last an ape to emulate.
Lots of sex: females with males; females with females; males with males; adults with children and infants...

—*Hold on a minute.*

A mother-son relationship so close that she's a guy's favourite wingman: she'll chase other guys out of the way so that her son can literally get the girl. No bonobo embarrassment about being tied to mother's apron strings here...

—Yikes.

No groups of males causing any trouble. No war. No bands of brothers. No male teamwork. No male team sports. No football...

—Now, that really is going too far!

It is not that some less than appealing bonobo features such as these are unknown, only that they have been largely avoided or ignored. When it comes to the extremely delicate issue of sex involving children and infants, this is not on the public agenda of most wannabe bonobos. Those who think differently will, as we'll see, not get the kind of support they might be expecting from the bonobo anyway. But sex involving infants and juveniles (youngsters between infancy and puberty) is an important aspect of bonobo life, and not something to which we can easily turn a blind eye.

As for mothers and apron strings, and guys and their team sports, these are probably the biggest blocks to the release of the inner bonobo for the males amongst us. Is the promise of rampant sex enough to compensate? Or is the male wannabe bonobo expecting to have his cake and eat it too? As for the gals, it would all seem something of a feminist paradise. We'll see if it is.

We are, as zoologist Desmond Morris pointed out to us back in the 1960s, "naked apes" (Morris 1967). But are we, as some now think, naked *bonobos*? Before we jump in, genitals first, perhaps we need to take a step back and consider the wider picture of bonobo behaviour. Might there be some aspects of being a bonobo that are not going to go down too well; aspects that require a little more scrutiny before we take the bait of 'peace and love' promises?

For the naked ape – and especially for the wannabe naked bonobo – it is time to take that step back and have a good look at this this new-found, rising-star, cousin of ours. It is time for the bonobo to lay herself bare.

One: the forgotten ape

In 1997, in the preface to his book *Bonobo: The Forgotten Ape*, Frans de Waal made the point that books and articles on the other apes could easily fill a small library, but for a complete collection of literature on bonobos, a single cardboard box would do. At that time few had even heard of this ape but now, nearly two decades later, the bonobo takes the spotlight in many a discussion about human nature. It seems that we are finally getting to know our peaceful, hairy cousin; a peace achieved primarily (so we are told) through constant, casual sex.

Bonobo: The Forgotten Ape is full of wonderfully endearing bonobo photographs (though the unnaturally reproduced redness of the lips and skin is a little unfortunate). The man who captured these images, wildlife photographer Frans Lanting, describes in the book how finding and then keeping up with bonobos in their natural forest habitat was no easy task. What's more, if he did find them, photographs of black animals in a dark forest weren't going to show us much anyway. As a result, most of the photographs in the book are either of bonobos at the artificial feeding sites, or they are photos of captive apes, used to "show close-ups and subtle expressions".

While it is inevitable that most of the images we have of bonobos are from zoos or from the cleared areas of feeding sites (where they are often seen clutching their sticks of sugarcane), it can leave us with a distorted image of this ape. We have to remember that the world they normally live in is a world of trees and, while they will feed and travel on the ground and through the undergrowth, they spend most of their time traveling, feeding, and sleeping high in the forest canopy. Bonobos are particularly agile in the trees; their long-legged, slender body – once thought to show a closer connection to our own – is now recognised as a body-type well-suited to their largely arboreal life (White 1992).

The very same problems Frans Lanting experienced when trying to find and then follow bonobos in their natural forest habitat, plus the fact that this region of Africa has seen some of the most harrowing and long-running human conflicts, is why, even today, we know relatively little about these apes. Most of what we do know still comes, like our images, from tiny captive populations or from bonobo behaviour at the artificial feeding sites.

Frans de Waal is probably the only writer on bonobos who is familiar to the vast majority of bonobo enthusiasts. His major interest has been reconciliation behaviours across primates rather than simply this one species but the results from his study of a small group of bonobos at San Diego Zoo have massively impacted how most of us see this ape. Zoo studies have been useful in highlighting behaviours particular to bonobos but their use, of course, is limited. (Imagine if some alien visitors to our planet killed some mothers for food, took and traded their infants, and that these infants, and their descendants over time, end up in various zoos on the alien planet. What would alien scientists learn about us? Life for a bonobo may be far less complex than our own but it is far more complex than is ever going to be found in a zoo enclosure.)

Though our literature on bonobos has increased since 1997, it would still fit into a (somewhat larger) cardboard box. For those of us feeling the rumblings of an inner bonobo on the verge of breaking loose, it would be foolish not to at least take a good look inside that box rather than leave its contents to the workings of the sometimes over-excited human imagination.

Our first discoveries about bonobo behaviour came from observations of very small zoo populations. Frans de Waal's early 1980s study of the bonobos in San Diego Zoo is the most detailed of these, and it is de Waal who has been responsible for most of our general understanding of this species. Not surprisingly, it has been the sexual behaviours of the bonobo that have grabbed our attention, and the value of de Waal's broader concern with reconciliation and peacekeeping behaviours across species usually loses out to the titillation of bonobo sex.

For the public, the overriding bonobo image is one of sexual orgies and little else. As much as this can be an extremely inviting image for the wannabe naked bonobo, we – the most intelligent of the apes – surely need to live up to our name and look beyond this image before jumping in. We'll start with a fairly detailed look at de Waal's bonobos.

There were a total of ten bonobos at San Diego Zoo when de Waal carried out his study in the winter of 1983-84. At the start of the observations the bonobos were living in three subgroups, two of which were merged during the study period. One subgroup comprised an adult male named Vernon (about 12 years of age), an adult female named Loretta (about 10 years of age), and Loretta's adolescent brother, Kalind (about 7 years of age).

The second subgroup comprised an adolescent male named Kevin (about 9 years of age), his sister Louise (about 12 years of age), and Louise's two-year-old female infant Lenore (fathered by Vernon).

The final subgroup comprised four juveniles: two of each sex, three of them full siblings, and all less than six years of age. The first and second subgroups are the ones that were merged.

Firstly, it should be noted that even the oldest bonobos here are still barely adults. How the different life-stages are defined can vary, so the 7-year-old male Kalind, described as adolescent by de Waal, might just as well be described as late-juvenile. Vernon, Loretta, and Louise are described as adults here but, with ages of 10 to 12 years they could just as well be described as late-adolescents.

In the wild, most bonobos are 14 or 15 years old before they produce offspring, and are usually defined as adults only when they are into their teens. At the time of the early studies of captive bonobos (in the 1970s and 1980s), it was not unusual for the oldest members of the bonobo groups to still be very young but it did lead to some misconceptions about the sexual behaviour of adults. We'll come back to these misconceptions later but, for now, we just need to note the young ages of the bonobos in this zoo group.

Secondly, it should be noted that apart from three bonobos – the (barely) adult male Vernon, the female infant Lenore, and one of the two male juveniles – all the other seven bonobos are full siblings. Their parents (Linda and Kakowet, the original pair of bonobos at the zoo) produced ten offspring over fourteen years, each one taken to be reared in the zoo nursery. Under natural conditions their mother would have been infertile for a number of years after giving birth (while she nursed her infant), and she would have only produced perhaps three or four offspring in that fourteen year period. Removing the babies from their mother enabled a very prolific reproduction rate.

More importantly, under natural conditions the sexually mature brothers and sisters would not be living together. In the wild, sexually mature male and female siblings live in different bonobo communities because, as with the chimpanzee, young adolescent females migrate to a new community where they will breed with unrelated males. So this zoo population does not replicate natural conditions on many levels but we'll leave these issues to one side and simply assume that these bonobos are otherwise showing us natural behaviours.

In this highly influential study we have ten bonobos in all: one adult male, two adult females, two adolescent males, one infant female, and four juveniles. These ten bonobos were observed for 300 hours, during which 698 "sociosexual encounters" were recorded, which is an average of more than two sexual interactions per hour across these ten bonobos. That these "sociosexual encounters" include oral sex, "kissing with tongues", and the massage of another's genitals, not to mention a lot of homosexuality, it is no surprise that this veritable smorgasbord of sexual acts can leave people thinking "orgies". More than this, some also think that such behaviour is a mirror reflecting our own natural selves. Is it? We need more details to help flesh out the images dancing in our heads.

Between the adult male and the two adult females there were only 31 sexual encounters, which is about one every ten observation-hours. Prior to the merging of the two subgroups there was only one adult female with the adult male but still, this seems relatively slow-going for the adults. The two adolescent males, though, did a lot better: there were 168 sexual interactions between the adolescent males and the adult females. Most of these were, as with the adult male, face-to-face (ventro-ventral, in the more scientific lingo).

What about homosexual encounters? The numbers for homosexual encounters amongst the adults and the adolescents are:

86 sexual encounters between the adult male and an adolescent male,

17 between the two adolescent males, and

65 between the two adult females.

In total then, we have 367 sexual encounters amongst the adults and adolescents: 199 of them heterosexual, 168 of them homosexual.

Another 76 sexual encounters were recorded amongst the juveniles (who, remember, were in a separate group from the other bonobos for the whole study period). Of these:

46 sexual encounters were between the two males, and

29 were between a male and a female.

Between the two juvenile females there was just one solitary 'sexual' act: an open-mouth kiss.

We are still left with 255 sexual encounters to account for, and it may come as something of a shock to discover just who was involved in all of these sexual encounters: the female infant, Lenore. She managed to amass:

196 sexual encounters with the adolescent males,

47 with the adult male, and

12 with the adult females.

That's quite some going. About a third of these were initiated by the infant herself: often she would climb onto the belly of an adolescent male and press her genitals against his, and the adolescent male would respond with some pelvic thrusting movements. At other times she would present to the male for a ventro-dorsal mount (i.e., doggy-style).

Is this quite what we had pictured in our heads? More crucially, do we still think we are looking in the mirror?

We'll put the antics of this young female to one side for a moment while we consider some of the bonobo sexual behaviours that get more attention when it comes to attempts to draw parallels between ourselves and the bonobo.

Overall, 420 of the sexual encounters were face-to-face. This is a sexual position once presumed to be unique to humans, though there is some confusion as to whether this is a naturally preferred position or one imposed by religion and missionaries, and therefore, perhaps, to be spurned. For these bonobos, 156 of their sexual encounters were in the more animalistic doggy-style position, while 23 involved the parties facing in opposite directions while rubbing their rears together. How often humans engage in the latter position will have to be left to the imagination.

Kissing is certainly quite prominent in the human repertoire, and there were 43 cases of an open-mouth kiss amongst these bonobos, with about one quarter of them involving tongue-tongue contact. Kisses were included as "sociosexual encounters" by de Waal because of what he described as "their erotic nature".

So, who was doing the kissing? 34 of the 43 kisses were between juveniles: 20 between two males, 13 between a male and a female, and 1 between the two females (the only "sociosexual encounter" of any kind between these two).

Of the remaining 9 kisses, 5 were between the adult male and an adolescent male, 2 were between the two adult females, 1 was between an adult female and an adolescent male, and 1 was between the infant and an adolescent male.

Whether or not we are surprised that it was the juveniles who were responsible for most of the "erotic" kissing, it is surely more than a little disappointing that not a single kiss occurred between the adult male and either of the adult females.

What about oral sex?

There were 17 encounters involving oral sex (all fellatio), and all but one of these occurred between juveniles: 8 between the two juvenile males, and 8 between a juvenile male and a juvenile female. These cases all occurred during play when the young males briefly thrust their penis in the mouth of a playmate. The one case that did not involve juveniles was between the adult male and an adolescent male. So, not a single case of "oral sex" (and even that term can only be applied loosely) involved sexually mature males and females.

Finally we come to the genital massage. There were 39 sexual encounters where an individual briefly massaged the genitalia of a male. This time the juveniles are only responsible for 2 cases: 1 between two males and 1 between a male and a female. The adult females were involved in only 3 cases: 2 of these with an adolescent male and 1 with the adult male. The remaining 34 episodes of genital massage were between the adult male and one or other of the two adolescent males. These were instances of conflict resolution, occurring after the older male had chased the adolescent, and the latter would present his penis to the adult.

We'll be looking at homosexual behaviour of both sexes later but for now, what more can we learn about the "sociosexual encounters" amongst the three sexually mature (adult and adolescent) males here? Is their "make love not war" behaviour showing us how they have overcome sexual competition by simply enjoying a bisexual free-for-all?

It turns out that the adult male was not at all happy with the sexual advances of the adolescent males towards the adult females, regularly stopping them in their tracks. When it was only the 7-year-old male in the same group as the adult male this was not so much of a problem; at only 7 years of age the male was still hardly adolescent, and juvenile male behaviour (as we'll see) is generally highly tolerated by adults. But when the two subgroups were merged it meant that the older adolescent male was now in the same group as the adult male.

Initially, the meeting between the adult male and this older adolescent had appeared to go quite well, with no serious aggression, but by the second day the adolescent male had a canine gash on his upper lip. This injury was enough to put an end to the adolescent's sexual advances towards the females, at least when in view of the older male.

In fact, after the study period ended, and after months of mounting tension between these male bonobos, it was decided to remove both adolescent males and put them in the juvenile group. This all comes as something of a blow to the species' sexually relaxed, "make love not war" image.

While it takes some time and effort to sort out who was doing what to whom in the San Diego bonobo groups, it is an important exercise when attempting to locate our inner bonobo. Is this what we imagine for ourselves as naked bonobos? Is this even what we'd imagined for hairy bonobos?

Not surprisingly, the sexual spotlight has focused on the face-to-face sex (suddenly the 'missionary position' gets the bonobo stamp of approval), the homosexuality, the oral sex, and sometimes the kissing. Yet 255 out of 698 incidences of sex, that's 36.5%, involved the infant female. A further 76 sexual encounters involved the juveniles, which means that over 47% of the zoo sex involved the sexual immatures. Add to this the fact that out of all the sexual encounters it was the juveniles doing most of the kissing and virtually all of the oral sex, and noting the relatively low-key nature of heterosexual activity between sexually mature individuals, is this really, as de Waal tells us, Kama Sutra sex?[1]

The most important point to make here, and one that may not have been made crystal clear to those who think they know about all this bonobo sex, is that the sexual behaviours involving immatures do not involve intromission (penetration) or ejaculation. Ejaculation only occurred (though did not necessarily always occur) during copulation with mature females; in this way, de Waal says, the (potentially) reproductive function of sex is kept separate from the social function. It is an interesting distinction.

It is useful to note here that, as well as all the sociosexual behaviours listed above, there were also 39 observations of solitary masturbation (most of these by the adolescent males, some by the adult females, some by the juveniles, and none by the adult male) and again, the males did not ejaculate; probably not something the male wannabe bonobo has had in mind.

---

[1] Frans de Waal repeatedly calls bonobos the Kama Sutra primates (e.g., de Waal 1989, 1997, 2005).

The naturally restrained nature of bonobo sex when it occurs between sexually immature and sexually mature bonobos is surely some relief. At the same time, it might be disappointing to discover the rather tame nature of so much of the sexual, especially heterosexual behaviour amongst the adolescent and adult bonobos. One thing, though, is certain: if we are going to call all these bonobo behaviours "sex", then we also need to be clear about who is actually involved and what actually occurs.

When we learn that bonobos have sex all the time we probably picture a bonobo scenario where a lot of random sexual activity goes on, and if sperm happens to meet egg and a female becomes pregnant, then that is incidental. It is the idea that bonobo sex is a recreational activity totally disconnected from considerations – conscious or not – about reproduction. But such an umbrella term as "sex" has the potential to do more harm than good when it is left up to the imagination as to what any particular sexual encounter might entail.

When the word "sex" is applied to a multitude of different behaviours simply because genitals are involved, we are easily misled into thinking these behaviours are more erotic in character than they actually are. Some are not erotic at all. Field researchers tend to use terms such as "genital contacts" and "pseudo-copulations" for many of these behaviours – terms too quickly mocked by those who mistakenly think this is an attempt to protect sensitive readers from shockingly unbridled sexual behaviour in another species. It isn't.

Using the term "sociosexual behaviour" may seem better than using the term "sex" but then, do we use this term for all these behaviours? Or do we exclude heterosexual copulation between a mature male and fertile female because that is, potentially, reproductive sex? These San Diego Zoo bonobos kept penetration and ejaculation for interactions that might possibly lead to conception, and clearly there was competition between the males over access to fertile-looking females, so they would appear to know when they are potentially having reproductive sex rather than social sex.

If we use "sociosexual behaviour" to encompass all of the bonobo behaviours then we have the same problem as using the term "sex", and we are using it for any genital contact, whether it is one between a playful infant and an adult male, one between an infant and its own mother, a barely erotic 'oral sex' encounter between juveniles during play, a dominance-based mount by an adult male, and so on, up to a full heterosexual adult copulation. For such a broad spectrum of behaviours to come under a single word or term obscures important details – details that matter if we are not to be misled about what actually occurs.

Much of the bonobo sexual activity is not about reproduction but it is often not recreational either; it is used to manage social conflict and social stress. Does this mirror our own use of non-reproductive sex?

When sex is used primarily as a way to manage social conflict and stress, then it is not something that can be limited to a private activity between sexually mature individuals who find each other sexually attractive. Using sex for social communication in the way of the bonobo requires a vast expansion of such activity beyond particular, desirable partners. Do we really have an inner bonobo able to make genital contacts with, well, with *anyone*? Or are we "only when it involves sex with those we find attractive" wannabe bonobos? Can we each take just a piece of 'bonobo' that we like, and simply ignore the rest? Not if this release of our "inner bonobo" is meant to be taken seriously.

If we are really seeking peace through sex, then that sex will have to occur with those we don't actually like; those we would otherwise be in competition and conflict with. But we are jumping ahead already, and really need to learn a lot more about bonobos before we can properly start to think about this, and what it is we are really hoping to find within ourselves.

In our quest to get to know as much as we can about the bonobo, the details from de Waal's study have given us somewhere to start, and they present us with many questions about bonobos, about ourselves, and about sex. But it is a captive study, and we need to know much more than these zoo exhibits can ever hope to tell us. It is time to go wild.

The first thing to remember about bonobos is that they are extremely difficult to observe in their natural habitat. Deep in the rainforest, south of the Congo River, they are, in the first place, difficult to even find. Then, if chanced upon, they will flee rapidly through the forest canopy or the dense forest undergrowth. Getting used to human observers takes years for this nervous, highly alert ape, and the females can be particularly reluctant study subjects.

In 1972, American anthropologist Arthur Horn attempted the very first bonobo field project but in a two year period managed a mere six hours of observations. That gives some idea of just how difficult it has been to find and study these animals. Fortunately, in 1973 the Wamba and Lomako study sites were established, and gradually data on these secretive apes could be collected.

At Wamba it was decided to create artificial feeding sites to coax the bonobos out of the forest and make observations easier. Consequently, most of our information from the wild comes from Wamba, and most of that comes from bonobo behaviour at the main feeding site where trees were felled, and food — mainly sugarcane — was laid out for them. Researchers were then able to watch the bonobos from an observation hut built upon a large old termite mound.

To follow on from what we learned of the San Diego Zoo population studied by de Waal, it would be useful to first know more about the sexual behaviour of bonobo youngsters in the wild. While literature on the bonobos is strewn with observations of sexual behaviour, there is very little directly concerning that of sexual immatures. The main source for such information is a study of bonobo sexual development by Chie Hashimoto (1994, 1997). This is a study of the E1 group at Wamba over the period from November 1990 to February 1991, and is based on about 100 hours of observations, mostly taking place at the artificial feeding sites.

E1 is the most familiar of the bonobo communities at Wamba, and at the time of this study comprised only 31 members who tended to range as a single party, except for one adult male who ranged alone. Bonobos, like chimpanzees, have what is called a fission-fusion social organisation which means that members separate into small foraging parties, with individuals leaving and joining different parties for varying amounts of time. All members of these foraging parties, though, belong to a single group which is known as the unit-group or, more usually now, the community. Males stay in the same community for life but females transfer to a new community, usually around the age of 8-10 years. E1 is quite unusual in that nearly all members tended to be together at the same time, probably due to the relatively small size of this community, and to the availability of large food sources, including the provisioned sugarcane.

So, what did Hashimoto find?

Hashimoto found that infants less than one year old were already engaging in genital contacts. Amongst all the sexual immatures (infants and juveniles), genital contact mostly occurred during play behaviour, and it occurred most frequently between two males; young bonobo males, like males of most species, tend to play more than the females. Genital contacts were also more frequent between siblings, probably because they are the most readily available playmates. When infants were less than two years of age the genital contacts were face-to-face but older infants and juveniles engaged in some doggy-style mounts. The frequency of the doggy-style mounts increased with age, though females were only ever in the role of mountee.

So from a very young age, bonobos, especially the males, are engaging in sex with their peers, making brief thrusting movements with each other during their rough-and-tumble play.

Hashimoto also recorded 19 cases of genital contacts between mature (adolescent and adult) males and the immatures. Of these, 6 involved an infant and mostly occurred during play when the infant was held belly-to-belly, and the mature male shook the infant's body so their genitals rubbed together. We therefore have far fewer of these types of sexual encounters than little Lenore got up to at San Diego Zoo, but they do, nevertheless, occur in the wild.

In contrast to the playful nature of these interactions with infants, the 13 situations where the immatures were 4 years old and over were quite different. These genital contacts were not playful but occurred after the adult male had attacked or threatened the immature who then presented to the adult and was briefly mounted by him. Unlike the play behaviour with the infants, this was clearly appeasement behaviour by the juveniles, and it was used to end a conflict situation.

Interestingly, 7 of these 13 incidents involved a juvenile female whose mother had recently died. Without a mother to look out for her, this young female was often on the receiving end of aggression from others, and therefore had to show frequent appeasement behaviour. This aggressive treatment of a motherless young female is probably not what most of us would expect from the bonobo, and it does nothing for their 'one big happy family' image.

Sexual behaviour between immatures and adult females (there were no adolescent females in the group) was quite different again. There was a single observation of an adult female making genital contact with a 4-year-old female but there were 71 observations of genital contact between adult females and immature males. Of these, 10 were between a mother and her infant son, and occurred when the mother was stressed during tense situations. It seems that rubbing her infant son's genitals against her own acted to reduce the mother's emotional agitation. Considering how much of a struggle it is for modern women to breastfeed in public, this is probably not to be recommended for stressed-out, naked bonobo mothers.

Males older than 2 years of age were keen to engage in genital contacts with adult females other than their mother, and only 2 cases of mother-son genital contact were seen after this age. When a female was engaged in adult copulation the young males often eagerly approached and joined in with their own mating attempts. From 4 years of age the sexual behaviour of these immature males was much like that of adult males, including soliciting behaviour towards the females.

As males approached adolescence there was an abrupt drop in frequency of their sexual behaviour. The sexual behaviour of immatures had been highly tolerated by the adults; that of adolescent males was not. As they approached and entered adolescence, young males were persistently threatened and attacked by the adult males, forcing them to the periphery of the group.

This aggressive treatment of adolescent males accords well with what transpired in the San Diego Zoo group. Persistent threats and attacks from bigger, stronger, adult males towards these younger males with their budding reproductive capability is behaviour that does not sit well with our stories of a relaxed, sexual free-for-all.

As the adolescent E1 males moved into adulthood their frequency of genital contacts increased but this was mostly with other adult males rather than with females, and occurred in the context of conflict resolution. The amount of sex each adult male had with the females varied, but Hashimoto found that not one of them was getting as much sex with adult females as they had got as juvenile males. Heterosexual sexual activity for these E1 males peaked, therefore, as juveniles; something for the wannabe bonobo to ponder.

Immature females, in stark contrast to the juvenile males, had a very low frequency of sexual behaviour. It is particularly noteworthy that sex (or at least, immature copulations without ejaculation) involving males between the ages of 2 and 6 years with adult females was so frequent whereas sex involving young females with adult males (or adult females) was so rare. It raises questions about the supposedly casual, all-encompassing, *social* nature of bonobo sex, revealing instead the *reproductively based* differences between the sexes in how and why they participate – or not – in sexual activity.

While it may be encouraging to note the relative lack of sexual interest adult male bonobos show for juvenile females, it is hard to imagine much enthusiasm from our own adult female wannabes when it comes to the accommodation of the sexual desires of immature, juvenile males.

For the E1 bonobos, the frequency of adult female sexual behaviour was higher than that for the adult males, most of this being genital contacts – gg-rubbing – with other adult females. This gg-rubbing is the distinctive female bonobo behaviour where (usually) one female stands on all fours and the other holds on from below, as if she is an infant being carried, and the two females rapidly rub their swellings together in a side-to-side motion. It is probably the most well-known of bonobo sexual behaviours, and we will give this a lot more attention later.

In her study at the artificial feeding sites, Hashimoto found rather more adult homosexual behaviour than heterosexual. Sexual behaviour between the adult males – predominantly a brief mount or a rump-rub – occurred clearly in the context of conflict and aggression. The rump-rub seems to be a consequence of both males presenting for a mount and so, because they are facing away from each other, they press their rears together rather than one of them mounting the other.

This study by Hashimoto is the only one we have which is primarily focused on the sexual behaviour of immature bonobos. Additional details of immature sexual behaviour are found within studies with a much broader or a quite different focus. One such study is by Kitamura (1989) who carried out a five month study of genital contacts amongst the Wamba bonobos in the late 1970s. Kitamura recorded 93 'copulatory' interactions between immature males and mature females, 20 of these involving an infant male. Mature males were involved in 49 episodes of genital contact with infants of both sexes, more or less evenly divided between face-to-face and 'doggy-style' encounters.

Kitamura's study was of the E community before it split into E1 and E2, and there were nearly 60 identified members including 19 infants. This higher number of individuals explains why there is the greater frequency of behaviours involving infants compared to the later study of E1. As in Hashimoto's study, juvenile females were rarely observed to be involved in any sexual behaviour.

Takayoshi Kano is the Japanese researcher who initiated and led the studies at Wamba, and he is the author of *The Last Ape* (English Translation 1992, originally published in Japanese in 1986), a comprehensive overview of the bonobos at Wamba. Kano's book is written for a popular audience and, thanks to its translation into English, it has been an important source of information for many westerners on wild, though artificially provisioned bonobos.

In this early book on the Wamba bonobos, Kano writes that the male juveniles vigorously pursue sexual interactions with adult females, and only occasionally (5 cases) were these seen to occur between mother and son. Male infants, he tells us, begin to show sexual behaviour at less than 1 year of age, and an infant will insert his penis in the female partner of his mother after their gg-rubbing. As juveniles, male bonobos will race to join in when they see adults copulating or gg-rubbing, and they will often be assisted in their copulatory efforts by the female. If these juveniles are ignored they scream, and the female may then, Kano writes, "lose patience and wearily lift her hips as if to say, 'he's just a hopeless child'."

Adult males are also, we are told, enthusiastic about the "sex education" of younger group members. After an adult copulation they will mount and thrust at either male or female juveniles that present to them. The adult male does not insert his penis but rubs against the top of the hips and thighs. An adult male may also take a juvenile from its mother and thrust against it while it is clinging to his belly, or he may approach and mount a solitary juvenile.

Kano also tells us that sexual behaviour involving immature females with adults of either sex was relatively infrequent: 31 cases for females compared to 227 for males (based on 330 observation-hours at the feeding site during 1978-79). Most of this juvenile female sexual behaviour with mature adults was with males, and there were only 7 cases of gg-rubbing with mature females.

For immatures of both sexes, Kano observed that there was more sexual behaviour with mature partners than amongst themselves. He concludes that the immatures do not need to get involved in sexual games with peers "because they have the good fortune of receiving direct coaching from experts on sex, the adults".

As much as humans have problems dealing with the sex education of our children, this is probably not what even the most enthusiastic wannabe bonobo has had in mind.

We also have some information from the early 1980s on the bonobos at Lomako (Thompson-Handler, Malenky, and Badrian 1984). This study site was also set up in 1973 but, unlike Wamba, the Lomako bonobos have never been provisioned, and at the time of this study they were still not habituated (i.e., accustomed to the presence of human observers). Most of these observations had to be made using binoculars, and they were only of bonobos in trees as they would flee if attempts were made to follow them on the ground.

Of 75 copulations (or copulation-like interactions) seen between males and females during 414 hours of observation over an 18 month period, 69 (92%) were between sexually mature individuals. Of the remaining 6 interactions, 4 were between infant males and adult females, 1 was between an infant male and an infant female, and 1 was between a juvenile male and an adult female. None, therefore, occurred between a sexually mature male and an immature female.

The authors of this study note how infants are often involved in adult sexual activity due to the simple fact of being held by their mother while she is mating. On two occasions female infants were seen to approach adult males and handle the male's genitals, and also to rub their clitoris against the male. The adult males showed little reaction to these infants and showed no sexual interest or arousal.

With regard to homosexual activity at Lomako, 6 out of 25 cases of gg-rubbing between two females involved immatures: 3 were between two infants during play, 2 were between an infant and her mother, and 1 was between an adolescent and a juvenile. There were only 5 cases of any form of male-male genital contact, and only 1 of these involved sexual immatures: two juveniles thrusting in a face-to-face position.

From these results it would seem that there was a lot less sexual activity involving immatures going on at Lomako than at the Wamba feeding sites. On the one hand, it is clear that youngsters are involved in sexual activity, including with adults. On the other hand, it is very difficult to ascertain the extent of any sexual behaviour for your average bonobo when results differ across different field studies.

Captive studies present even more problems when it comes to trying to figure out bonobo behaviours, especially as these populations are so small. The San Diego Zoo population was relatively large at ten bonobos, but they were separated into smaller subgroups and there was only one adult male. The juvenile males and females were in a separate subgroup of their own, so there wasn't even the opportunity to observe interactions between juveniles and sexually mature individuals. And the solitary infant, involved in 255 of the 698 sexual interactions, lived with adults and adolescents only. There were certainly far more cases of sexual behaviour between infant and matures in this captive group than in any of the field studies.

In the San Diego Zoo population there were 168 sexual encounters between the adolescent males and the adult females, in contrast to Hashimoto's findings at Wamba where adolescent males were largely excluded from sex. This can be explained by the fact that the older of the two captive adolescents was, for a while, in a subgroup separate from the adult male, and therefore this adolescent male's sexual antics could not be stopped by him. The other adolescent, who was in the subgroup with the adult male, was only 7 years old and therefore closer in age to an unthreatening juvenile than a sexual competitor. Things changed when the subgroups were merged, and then we saw aggressive behaviour in line with that at Wamba.

The two zoo subgroups had been merged in order to see what would happen between the males but the San Diego zookeepers certainly did not want the bonobo brothers and sisters to breed together. In the end, the adult male's increasing intolerance of the adolescents coincided with the zookeepers' need to avoid fertile sex and inbreeding. Neither the keepers nor the adult male bonobo would, though for very different reasons, allow the potentially fertile matings by the adolescent males.

When it comes to the Wamba studies, we also need to take into account that the artificial feeding site enabled particularly large parties of bonobos to feed together. It is the availability of abundant food that brings out the bonobo's agitated and excited behaviour, including display behaviours, conflict and aggression, and it is in this rather chaotic context that the majority of sexual behaviours were seen. We therefore need to keep in mind that more natural small parties in more natural foraging conditions will show different results than we get from these concentrated feeding observations with their high social tension.

Unfortunately, these few studies are all we have for information on the sexual behaviour of immatures. Until we get more information from more populations of bonobos we cannot assume observations in zoos, or even at artificial feeding sites, are showing us how bonobos generally behave.

There are no reports from field studies of oral sex or of genital massage for any age group, so it is possible that they are rare or even absent in the wild. Nevertheless, sociosexual, non-reproductive behaviour is a significant aspect of bonobo society, most clearly embodied in adult homosexual behaviour. We will soon be taking a closer look at adult sexual behaviour but first we need to try to reach some sort of closure on that of the immatures.

At this point it would be useful to see if anything similar to bonobo immature sexual behaviour occurs in the other African apes. What about chimpanzees and gorillas? What do their youngsters get up to? Before looking at the small amount of information we have from these other apes, it is important to keep in mind how difficult it is to compare data coming from different animals in different environments. Even comparing information from different populations of the same species (such as bonobos at Wamba and at Lomako) is difficult when environment and demography, as well as the particular focus of the researcher, can vary so much.

The best we can do with regard to gorillas is a study of wild mountain gorillas from 1981 (Nadler 1986), which reports 28 episodes of immature sexual behaviour during 220 hours of observations. All of these were in the context of play, and usually comprised two or three pelvic thrusts against the body of a playmate. The playmates on the receiving end of this thrusting were of both sexes but the thrusters were always male. There were, though, 4 episodes where one particular immature female pressed an infant against her genitals.

This gorilla study only briefly mentions sexual behaviour between adults and immatures, stating that there were about three times as many mounts of immatures by mature males than there were mounts between immatures, but there are no further details. It is also noted in this study that adult females in the wild have occasionally been observed to thrust their genitals together while in the face-to-face position.

For gorillas, therefore, we have some brief mounting amongst immatures during play, and we have rather more mounting behaviour directed by adult males towards immatures, most likely as a form of dominance behaviour. There is also the one immature female pressing an infant against her genitals, and the occasional observation of genital contact between adult females. Some of these behaviours amongst gorillas may raise a few eyebrows but they do not really equate with the level of such behaviours amongst bonobos.

We have more information on chimpanzees than we do on gorillas. Jane Goodall (1968) describes some of the immature sexual behaviour at Gombe. She writes that two infant males, Flint and Goblin, mounted and thrust on their mothers when they had sexual swellings. Flint's mother played with his penis when he was young, and she tickled her daughter Fifi's genitals. Goodall says that the sexual behaviour of the two infant males was almost fully developed by 1 year of age.

Flint made his first genital inspections – of his 5-year-old sister – in his fifth month of life. This was followed by genital inspections of almost all females that stopped near him, and occasionally of some of the males. At 6 months of age Flint had an intense interest in sexual swellings: touching, smelling, and repeatedly licking them. At 9 months of age he made 10 attempts in 15 minutes to mount a female. The female crouched low for him and he was able to press his erect penis against her swelling and make a few erratic thrusting movements.

As with the other apes, sexual behaviour amongst young chimpanzees is often a part of their play behaviour, briefly mounting and thrusting doggy-style or face-to-face. Again, it is the males who are most active in these behaviours, and, while males may be in either position during doggy-style mounts, the females are always the mountee, never the mounter. Juvenile male chimpanzees, like juvenile bonobos, were also seen to mount and thrust on receptive mature females.

Caroline Tutin also observed chimpanzee sexual behaviour during her Ph.D. research in the early to mid-1970s (Tutin and McGinnis 1981). She writes that juvenile males frequently mounted and thrusted against younger individuals during play, noting that this did not involve intromission. In 1200 hours of observations she also recorded 408 immature copulatory mounts by males (infants and juveniles aged 15 months to 9 years) of 9 adult females, and 318 of these mounts included intromission. As with bonobos, juvenile female chimpanzees were involved in few sexual encounters. Tutin writes that juvenile females from the age of 4 were seen to present to their peers during play, and to adults in greeting, but these juvenile female presentations were uncommon.

Sexual curiosity and activity is clearly a normal part of ape development, and in the case of chimpanzees and bonobos, mature females are highly involved in the sexual learning process for the sexually immature juvenile males. In fact, juvenile males of both species are involved in more sex with mature females than are adolescent or adult males. This may partly be because these younger males are not ejaculating, and so they can more readily repeat the behaviour than can a 'spent' sexually mature male, but there is also the greater general tolerance shown towards these young – and infertile – sexual learners.

Adolescent male chimpanzees do not appear to have the abrupt drop in sexual behaviour that hits the adolescent male bonobos. This is not because adult male chimpanzees continue to tolerate the sexual behaviour of these maturing males; it is more likely because chimpanzees are more dispersed, so adolescent males may have more opportunities to get close to receptive females without an adult male necessarily spotting him and getting in the way. Bonobo females, in contrast, are more clustered together in the centre of the foraging party, and adult males stay close to these females, scaring off the younger males to the edges of the grouping.

There are general similarities between bonobo and chimpanzee immatures in their sexual behaviour. In both species there is sexual behaviour from a very young age, often amongst peers during play but also with mature individuals. Immature males engage in much more sexual behaviour than do immature females, and these young males have frequent copulation-like (non-ejaculatory) interactions with adult females. Immature chimpanzees sometimes engage in face-to-face genital contacts with each other, while this position is the predominant one for immature bonobos.

Though we still have very little information on immature chimpanzee sexual behaviour in the wild we tend to assume, rightly or wrongly, that the frequency is somewhat lower than that for immature bonobos. We might reasonably expect the frequency to be higher in bonobos simply because bonobo youngsters are likely to have more opportunities to interact with others. This is because their mothers are more often in the company of other bonobos, rather than foraging alone as chimpanzee mothers are reported to do.

Interestingly, de Waal (1995) did carry out a comparison between the San Diego bonobos and an outdoor colony of chimpanzees at Yerkes, and he found the sociosexual behaviour rates of the immatures to be similar. Another captive study comparing the development of bonobos and chimpanzees found no difference between the two species in the rates of mounting behaviour by infants. This study also found that the infant male chimpanzees spent more time engaging in sexual inspections of other group members than did the bonobo infants (de Lathouwers and van Elsacker 2006).

We have discovered that young male chimpanzees have an active interest in sex with oestrous (sexually swollen) females, and engage in mounting behaviours with their peers during play. So, as long as playmates or oestrous females are available, perhaps the sexual behaviour of the immatures of both species is not so dissimilar. And if this is so, perhaps we need have no more concern about the sexual activity of immature bonobos than we have had about that of immature chimpanzees.

If both species are doing pretty much the same until sexual maturity, then what the infants and juveniles get up to would seem to be irrelevant in the call to release our inner bonobo. We can breathe a sigh of relief and say the immature sexual behaviour of bonobos is just the way so many other species learn and practise sex in preparation for adulthood; what bonobos do in this respect should then be no more of an issue for us than what the young of other species do.

On the other hand, if there are differences between immature bonobos and the other immature apes we need to consider how those differences might play a part, and maybe a crucial part, in the development of sociosexual behaviour as adults, and what this implies for wannabes. Does the widespread use of sociosexual behaviour as adults require the expression of sociosexual behaviour by immatures? And is such behaviour innate or learned?

There is a recent study, "Bonobo but not chimpanzee infants use socio-sexual contact with peers" (Woods and Hare 2010), that, as the title states, found a difference between the two species that suggests infant bonobos may have an innate predisposition to engage in sociosexual behaviour. So, what did this study entail, and what did it find?

This study was of orphaned bonobos and chimpanzees in their respective African ape sanctuaries, where the behaviour of 8 bonobo orphans was compared with that of 16 chimpanzee orphans. All orphans were aged 2 to 4 years, and were kept in peer groups so they had no adult ape influence. They were observed for a 30 minute period before feeding, for 30 minutes during feeding, and for 30 minutes after feeding, and there were 10 of these sessions for each species.

The study found that sexual interactions occurred amongst the bonobos but not amongst the chimpanzees, and occurred almost exclusively during the 30 minute feeding period.

The authors conclude that:

1. Bonobo sociosexual behaviour is species typical and develops without the need for learning (as there were no adult bonobos with the infants)

2. Bonobo sexual behaviour is used for social bonding or the relief of anxiety (as there was little sexual behaviour outside feeding)

3. Bonobo sexual behaviour is not tied to reproduction (it did not imitate reproductive sex in that it involved same-sex combinations as much as heterosexual, and various positions rather than just the standard doggy-style mount)

This does seem terribly at odds with where the evidence so far has been leading (i.e., that immature bonobo sexual behaviour is much like that of chimpanzees, and is about play and sexual practice). What can we make of it? There are a few points to be made about the authors' conclusions before we consider what the results of this study might be telling us.

Despite what the authors argue, this behaviour could be learned behaviour. The orphans had come to the sanctuary as 2-year-olds, not as newborns. Having spent up to 2 years with their mothers in a natural bonobo community, and from the evidence we have of sexual activity in their first year of life in the wild, it is highly likely they would have had some early direct experience of sexual behaviours with others. They would almost certainly have observed plenty of stress-related sexual activity amongst adults.

For both species at this age, sexual behaviour in more natural conditions occurs mainly during play, and not during the agitated time around food. In this respect, the complete absence of any playful sexual behaviour amongst the chimpanzee youngsters is unexpected, but it is also unexpected that the young bonobos engaged in their sexual behaviour during feeding rather than play.

The final conclusion that bonobo sexual behaviour is not tied to reproduction, while true, cannot follow from the observations of same-sex behaviour amongst these bonobo infants. This is because chimpanzee youngsters (as we have seen above) also engage in same-sex behaviours, and use face-to-face positions not used as adults, yet chimpanzee sexual behaviour is said to be only about reproduction.

While it needs to be taken into account that these infants of both species were in a completely unnatural environment after being brutally orphaned (they were victims of the bushmeat trade), and some of the study's conclusions are shaky, the reported behaviour of these infant bonobos is, nevertheless, intriguing.

Under more natural conditions the infant bonobos would not be the ones engaging in such behaviours around food; they would be with their mothers while she and other adults engaged in their stress-relieving, social sex. It is more than likely the infants will have observed this adult behaviour around food many times before coming to the sanctuary, and perhaps it surfaced in these orphans in the absence of adult group members. Infant bonobos would normally stay close to their mothers for comfort and protection while she deals with the stress of the occasion, and they would normally access their food via their mother rather than having to compete for it amongst peers.

The bulk of the evidence points to bonobo sociosexual behaviour developing only very gradually as bonobos mature, and then predominating in the adults. If there is some kind of innate predisposition, as the authors of this study propose, there is no evidence for anything resembling this form of stress-related bonobo behaviour in human infants – and it would be hard to keep a lid on such an innate predisposition to manage social conflict with a genital rub should it actually exist amongst groups of infants in our nursery schools, kindergartens, playgroups, and the like. We cannot do the above experiment with orphaned human infants but, if we could, how many of us would seriously predict human infant behaviour to be like that of the infant bonobos?

If, on the other hand, sociosexual behaviour is purely (or predominantly) a learned behaviour in the bonobo, it is learned from observation of adult sociosexual behaviour, and we are then left with the thorny question as to how the wannabe bonobo proposes such sociosexual conflict resolution behaviour might be observed by their own young; or would they still want it kept hidden from children, to be communicated to them at some age-appropriate moment?

Whether the "inner bonobo" we are being urged to release is the adults only version or includes all ages, there are issues that need to be addressed. If "adults only", then this instantly sets us apart from the bonobo, while the full "all ages" (and public) bonobo experience is completely at odds with cross-cultural human behaviour.

We have ended this chapter with an interesting study which (not surprisingly) shows some atypical behaviour from both the bonobo and chimpanzee orphans. It may be pointing to something of an innate readiness to engage the genitals in bonobos, especially when experiencing social stress, but it also promotes the idea of a greater polarisation of the two species than is supported by all other studies. Mostly, though, it shows just how difficult it is to find the clear and simple answers we seek.

Is there anything at least reasonably concrete that we can take from all these studies?

Immature male chimpanzees and bonobos both show copulation-like behaviour well before weaning; both have a strong interest in sexual activity with sexually mature females sporting their sexual swellings, and both have more frequent copulation-like behaviours with these mature females than adult males get to have actual copulations. Practising sexual behaviour appears to be an essential part of development, and seems to be particularly important for males rather than for females.

Both chimpanzee and bonobo male immatures are having a lot more sex than is really necessary for this learning to occur when we consider how other species manage with a lot less practice. This difference may have something to do with the relatively high frequency of adult sex in both species, and the high degree of sexual competition and/or sperm competition these males will face as adults.

Immature female bonobos and chimpanzees show a relatively low frequency of involvement in any kind of sexual behaviour. As they approach adolescence their job is to find a new community to join, and it is in that new community where their sexual interest and behaviour ignites.

These similarities point to the immature sexual behaviour of both species being primarily about practice for reproductive sex as adults. While this may raise questions about the sexual development of humans, our focus here is only on the contrasts between bonobos and chimpanzees, and whether there might be any aspects of immature bonobo sexual behaviour relevant to the proposed release of our "inner bonobo".

We found only a few differences between immature bonobos and chimpanzees, such as the very occasional instances of immature females engaging in gg-rubbing with mature females (a sociosexual behaviour which epitomises the bonobo), the occasions when mothers relieved their own stress by engaging in genital contact with their own offspring, and some playful interactions between adult males and infants of both sexes.

This chapter has been mostly about the sexual behaviour of immature bonobos and the problems this potentially throws up for those with aspirations to be naked bonobos. Though there is not enough information to reach a full understanding of sexual behaviour involving bonobo immatures, it should at least be clear how different it is from child sexual abuse in humans. One obvious difference is the absence of sexual interest in sexually immature bonobo females, while the immature juvenile males are in hot pursuit of the sexually accommodating adult females. Clear differences in sexual behaviour exist between male and female bonobos, and across bonobos of different ages and developmental stages.

What does all of this mean for the wannabe naked bonobo? When we were simply seeing ourselves as naked *apes*, and most of the information we had on our fellow apes was from chimpanzees, the sexual behaviour of immature chimpanzees (or other species) was simply never an issue. Perhaps, then, it should just be ignored by the wannabe bonobo? Should we just 'forget' these awkward aspects of the now unforgotten ape?

With sex taking the limelight in bonobos, especially as a social mechanism that creates a more peaceful existence, it is harder to restrict discussions of their sexual behaviour to adults only. This is why we need to know the details; we need to know exactly what we are dealing with here, and prevent any misunderstanding of bonobo "sex", especially one that assumes infants, juveniles, adolescents and adults, as well as males and females, are interchangeable participants in any type of sexual encounter.

Probably the most important thing to take from all this is that penetration and ejaculation only (though not necessarily) occurred with sexually mature females. Reproductive sex is, in this sense, separated from other sex. For bonobo males, even an erection is not necessarily a feature of their sociosexual behaviour. The wannabe bonobo who has come to believe that this ape is genuinely engaging in 'Kama Sutra' sex, or that there is a bonobo thumbs up for pretty much anything imaginable in the realm of human sexual behaviour, has been led astray. Even oral sex boils down to little more than a brief thrust of the juvenile penis in the mouth of a playmate, and any other bodily penetration is restricted to the sexually mature vagina.

Regardless of how the immature sexual behaviour of bonobos compares with that of the chimpanzee, the sexual behaviour of the two species diverges as they enter adulthood. Only adult bonobos engage in a significant amount of sexual behaviour which is clearly not reproductive sex, and is clearly unlike the sociosexual behaviour of other primates: gg-rubbing in females, and (the far less frequent) rump-rump contacts in males.

For those who are simply looking for an inner bonobo who can show us the route to peace through the pleasure of adult sex, reality, though, still has to be faced. Conflict does not only occur between those who find each other attractive, and might also find sex – in private – an agreeable way to reconcile. For bonobos, public genital contact, potentially with any other individual in the group, is no problem; for humans, it obviously is – and even when we're only talking adults.

From the sexual behaviours of immatures and all the accompanying difficulties we need to face, we now move to the other end of the spectrum and look at what has been revealed about the heterosexual activity of the sexually mature bonobo.

Two: heterosexual sex

When something new comes along, our attention is inevitably captured by the apparent novel features of the newcomer. This is what happened when the bonobo made her way to centre stage after decades in the shadow of chimpanzee stardom. As time has gone by, though, and as more has been learned about different populations of both bonobos and chimpanzees in the wild, those highly promising novel bonobo features have become much less clear and certain.

Initially, the bonobo's more gracile body was thought to be more human-like; now it is recognised as a body more adaptively agile for a largely arboreal life. The belief that the bonobo was shining a more accurate light on our own ancestors' bodies than the chimpanzee had done has all but disappeared. What has remained, though, and what has recently exploded in the public imagination, is a belief in a particularly close relationship between bonobo and human sociosexual behaviour. From face-to-face mating, to all kinds of non-reproductive sexual activity, to a general peace-loving nature, the bonobo is presented as being a closer cousin than the chimpanzee.

Many of the early beliefs about the sociosexual behaviour of bonobos have become fixed in our minds, and any suggestion that things are neither as simple nor as rosy as first thought will likely be counterpunched with accusations of political or religious backlash. Is there really some kind of conspiracy against the bonobo; some attempt to tarnish their "make love not war" image? Or are we just becoming better acquainted with both species?

As we leave the immatures behind and move on to the adults, we'll just remind ourselves what has been revealed so far. We have discovered that the juvenile females have little involvement in bonobo sexual activity (are they being antisocial?), while the juvenile males are getting lots of sex with the sexually mature females – probably not what the wannabe bonobo gals have in mind. This juvenile sexual behaviour is not so different from that found in the chimpanzee, and seems to be practice for reproductive sex rather than the use of sex for social communication.

At the same time, we do get a sense that from infancy the bonobo is far more primed for genital-contact behaviours. Whether or not there is some innate difference between the two species, bonobos from their earliest months are observing, even if not participating in, the sociosexual behaviour of adults.

As the bonobo male reaches adolescence his sexual behaviour is no longer tolerated, and the adult males will likely chase him off to the periphery of the group. Juvenile females have been very quiet in the story so far, and as they approach and enter adolescence they too are to be found on the periphery of the group. In contrast to the males though, this is by choice, for their futures lie elsewhere. From about 7 or 8 years of age the young females will be visiting other bonobo communities where they will be interacting with bonobo strangers. This is when the bonobo female's sexual career is jump-started, and eventually she will make one of these visited communities her home for the rest of her life.

Homosexual behaviour plays a big part in the life of immigrant, adolescent females but in this chapter we are focusing on heterosexual sex amongst all the sexually mature (adolescent and adult) bonobos. The first question is: just how much sex does a sexually mature bonobo actually get?

When it comes to people, we can ask this question and (though the answer may not always be honest) get a response of so many times a week (or whatever period). Bonobos cannot lie but neither can they tell us anything at all. When we want to know just how often bonobos are having sex we have to rely on humans seeing and recording their copulations. No one observes these apes 24/7, so the copulation rates we end up with come from snapshots of their lives, and this can lead to distortions and misconceptions about their sexual activity.

Frans de Waal recorded 199 heterosexual sexual encounters between the three sexually mature males and the two sexually mature females in his San Diego Zoo study. If we take out the one kiss and the three cases of genital massage (remember, there was no oral sex), that leaves us with 195 encounters that were copulatory mounts. Taken at face value, if we work out the rate from the 300 hours of observations we get a rough average of one copulatory mount every 5 hours for each of the three males, and one every 3 hours for each of the two females.

But, of course, it wasn't 300 hours of continuous observation; these observations took place, de Waal says, over the winter months. If we make a reasonable assumption that those 300 observation hours break down into 60 days of 5 hours each, we then have an average of 3.25 copulations per observation day for the whole group. From a starting figure of 195 copulations we now have a figure for a bonobo male of once a day.

Going through actual numbers can feel like a chore, so we often make do with words such as "frequently", "often", "sometimes", "rarely", etc. But these words can mean different things to different people, especially when it comes to sex. Remember Woody Allen's *Annie Hall*? In the split-screen scene when the therapist asks "Alvy Singer" how often he and "Annie" have sex, his reply is a despondent:

"Hardly ever. Maybe three times a week."

When "Annie Hall" is asked the same question by her therapist she responds with the complaint:

"Constantly. I'd say three times a week."

Compared to three times a week, the once-a-day frequency for the captive male bonobos may seem a lot better, but it is not as frequent as some might have imagined for this big-balled male. (Bonobo testes are about four times the weight of human testes, and something the male wannabe bonobo can only dream about.)

Hold on though, aren't these captive bonobos also having a lot of sex when they are not being observed and recorded?

Probably not. Bonobos in zoos are not doing much of anything for most of the time. The highlight of their day is when food arrives, which is also when they have sex. It is unlikely that de Waal's zoo study excluded this window of opportunity for these sexy observations.

In a 1983 study of bonobos at the Yerkes Regional Primate Research Center, observations led to the conclusion that sexual behaviour occurred only when the animals were fed. This bonobo group comprised one adult male and three adult females, and sexual behaviour occurred only during the noon feeding sessions; no sexual behaviour was observed at any other time of the day from 9:00 a.m. to 4:00 p.m. The bonobos even refrained from any sex in 2 of the 17 observed feeding sessions, while 32 copulations were recorded in the remaining 15 feeding sessions (Blount 1990).

Clearly, it would be possible to observe bonobos for quite some time without seeing any sexual behaviour; alternatively, if they are only observed when feeding then their copulation rate can appear very high. Nevertheless, one or two copulations a day is not bad going. What about bonobos in the wild?

When wild bonobos (and chimpanzees) come across a rich food source they are enormously excited and agitated and, amongst other behaviours, they have sex. The early results from the wild were from the Wamba bonobos, and they were mainly from the artificial feeding sites where the provisioning of sugarcane and other delights meant that lots of bonobos could be seen together, and there was lots of sex.

We'll now return to Kano's book, *The Last Ape* (Kano 1992), to see if we can glean any useful figures from the early research at Wamba. During the 40 month period from October 1975 to February 1979, bonobos were observed in the forest for a total of 700 hours. During this time 219 copulations were recorded, or one copulation every 3 hours and 12 minutes of observation. But this figure is just the total number of copulations for any and all bonobos seen during these observations, and tells us very little.

The fission-fusion nature of bonobo communities, along with the difficulties faced by observers trying to follow bonobos in the forest, makes getting a clearer picture of bonobos in their natural habitat difficult. We can picture the researchers following and watching a foraging party for a brief period, and recording copulations when observed. The Wamba bonobos were often in quite large foraging parties, so only seeing one copulation every 3 hours or so sounds like a relatively rare event but we are still not really any the wiser when it comes to finding out how often a bonobo has sex.

On a general level, Kano writes that little copulation occurs in small foraging parties of less than 10 individuals but in larger parties things are different. When larger groupings find a rich food source, or when they hear vocalisations from other communities, or when two parties reunite, then there is sex. These are the anxiety-inducing occasions that bring out most of the bonobo's sexual activity.

In all these situations there is great excitement and, along with the sex, there is male charging display behaviour, branch-dragging, and aggression. The males often react to these occasions with their highly visible erections (interestingly, the non-erect penis of the bonobo and the chimpanzee is difficult to see as it is mostly retracted inside the body). Copulations, Kano explains, are in tension-filled situations and seem unnecessary when in a relaxed state. It appears that our cosy human snuggles in bed must be leaving our inner bonobo cold.

Is the bonobo's rather chaotic context for sex what the wannabe bonobo has in mind? Most bonobo sex is a way of dealing with social tension, so it requires the trigger of these anxiety-inducing situations. "Make love not war" is not about having a lot of chilled-out sex with particular partners of your choice, which then leads to a constant state of relaxation and feelings of goodwill towards all others. It is about being freaked out by one tension-ridden social situation or another, racing around screeching in panic, and some fairly frantic rubbing of the genitals to help calm the situation.

Kano also reports on 330 hours of observations of the bonobos at the feeding site from the first 3 months of 1978 and the first 2 months of 1979. During these observations, 515 copulations were recorded, or one copulation every 39 minutes. This is five times the rate of that recorded in the forest, which comes as no surprise considering this is a concentrated food situation visited by a lot of bonobos (about 20 sexually mature males plus a greater number of females).

The whole point of the feeding sites was to enable easier observations of bonobos but the information thus gathered is only telling us about the behaviour of large parties of bonobos when they have come together to grab a share of a highly desirable free meal.

Many of us will have seen video footage of the bonobos at this feeding site at Wamba: bonobos clutching pieces of sugarcane while taking part in an apparent orgy. Most of this sex occurs during the chaos of the first ten minutes or so on arrival at the site rather than continuing for the whole of their stay. But, leaving this particular point to one side, it at least looks like everyone gets to join in, don't they? Kano provides us with more of the details.

Contrary to the hopes and dreams of many a male wannabe bonobo, the frequency of male copulation, Kano tells us, depends on age and rank, and we are obviously looking at a less-than-equal access to females. Kano provides the hourly rates for the males: five high-ranking males averaged 0.27 copulations per hour (range 0.16-0.37), nine middle-ranking males averaged 0.06 copulations per hour (range 0-0.28), and five adolescent males averaged 0.13 copulations per hour (range 0-0.26).

As pointed out earlier when we considered hourly rates in captivity, these figures are not telling us that these bonobos are copulating throughout the day at this rate; it only tells us that, for example, a high-ranking male copulated about once for every 4 hours at the feeding site, a middle-ranking male about once for every 17 hours at the feeding site, and so on. These time periods will be made up from multiple visits to the site so there will be many sex-less visits too.

Of the nine middle-ranking males, the three lowest ranking did not copulate at all, and neither did one of the adolescents. Kano provides the total number of copulations for each individual male visiting the site in these 5 months of observations. The top-ranking male scored 81 copulations, the second-ranking scored the highest figure of 113 copulations, and 2 other males had scores in the 70s. Then we have 2 males with scores in the 40s, 2 in the 20s, and the remaining 11 had totals ranging from 0 to 15.

Though we still cannot tell how much sex the average male bonobo gets per week, these figures at least show that there is a lot of variation, and they also show that rank matters amongst bonobo males. Even at this feeding site, which brings out the most concentrated sexual activity in this sexy ape, it is not a free-for-all.

In general, the lower the rank of a male that came to the feeding site, the shorter was his stay. Most low-ranking males simply grabbed some sugarcane and fled to a safe place to eat it. Why did low-ranking males not hang out with all the others? Hanging out at the feeding site would mean being attacked by high-ranking bonobos; hardly the egalitarian, "peace and love" image we expect.

Many females stayed at the feeding site for a long time, and they mated with the remaining high-ranking males. Some low-ranking males did get to stay, and we can see from the figures given that the average copulation rate for the adolescent males was not as low as we might have expected. Though Kano does not explain why some lower-ranking males did get to stay at the site, it most likely had something to do with the presence or not of their mothers, and the protection these females can provide their sons.

So, were bonobo males having a lot of sex at the feeding site? It depends. It depends on which bonobo, and it depends on how long the other bonobos allowed him to stay near the females. It also depends on what we mean by "a lot".

What about the females?

In this study the females, unlike the males, are not identified as individuals but are classed according to maternal status, i.e., whether or not they have offspring, and then classed by the age of any offspring. As this status will have changed in the two study periods which are a year apart, we are only given a more general indication of the frequency of sex in these different maternal status categories.

The frequency of sex for the female bonobo depends, Kano tells us, on age and reproductive state. A female's "reproductive state" means her maternal status and the size of her sexual swelling, as almost all copulation occurred when the sexual swellings were near to or at their maximum.

When it comes to the influence of age on copulation rate, things are very different from the males as it is the adolescent females that have the highest frequency of sex. These young females were involved in 272 out of the 515 copulations observed at the feeding site. Mothers with 2- to 3-year-old offspring were the next most frequent females engaged in sexual activity, while females with infants less than 2 years old had the lowest frequency of sex.

It is the adolescent females who have joined the community (not those born in the community) that have the highest frequency of sex, though a long period of adolescent sterility means they will not conceive for 5 to 6 years (i.e., not until they are about 14 years of age). Adolescent females have irregular swelling cycles, and though their swellings are relatively small they are visible most of the time. This makes them almost continuously sexually attractive to the males, and they are equally almost continuously sexually receptive. Being so sexy is how young female bonobos (and chimpanzees) manage the difficult transition to a new community; it is ultimately a way of getting close to unfamiliar others – and, most importantly, closer to the food.

Kano writes that when these young females first entered the feeding site they looked for signs of males. When a male arrived at the feeding site his penis would usually become erect, and though the young females were hesitant about approaching males they would respond to even the slightest hint of courtship display by a male. After responding to male sexual interest and soliciting sex from all males at the site, the young females would then settle down to feed. These females also often made strategic use of sex: they would either present for copulation and then take the male's sugarcane, or take a piece of food and then present to the male, as if catching him off-guard (Kano 1992).

So, how much sex does a female bonobo have? Again, it depends on which female, or class of female. Young immigrant females have the highest frequency, though the maximum individual count was 69 within this class, which is lower than the number of copulations for the two top-ranking males (81 and 113 respectively). The maximum individual count among mothers with 2- to 3-year-old offspring was 63, though the

58

average rate of sex for this group was less than half that of the adolescents. Other classes of female did not contain any individual who mated more than 22 times, and their average rate of sex was much lower still. There were also some females who did not have any sex at all.

Immigrant adolescent females are undoubtedly the bonobo females who are the most sexually active, and they are the likely stars of the feeding site videos and photographs. Established bonobo mothers with status, on the other hand, can access food without necessarily having to use sex to appease the males. In effect, an increase in female social status leads to a reduction in her frequency of sex – another aspect of bonobo life that the wannabe naked bonobo might want to ponder.

These early research results come mainly from a Wamba bonobo community labelled the E group. This community was mostly seen to range as two distinct subgroups, only sometimes coming together as one, but in the early 1980s these two subgroups separated permanently and became known as E1 and E2. The E1 community was subsequently the focus of most research at Wamba.

Does any later research give us a better idea of how much sex the bonobo gets? A study of E1 community covered 97 days during the period August 1985 to January 1986 (Furuichi 1987). In 529 hours of observations – only 12% of them at the provisioning site – 109 copulations were recorded.

This community tended to range together as a single party, and comprised 10 sexually mature females (7 adult, 3 adolescent) and 9 sexually mature males (7 adult, 2 adolescent). This total of 109 copulations over 97 days is hardly more than an average copulation rate of once a day amongst all of these bonobos. Not only is this not an orgy, it is hardly any sex at all. As 88% of observations took place away from the feeding site we can see the difference the feeding site made to bonobo sexual behaviour.

What about the bonobos at Lomako? We have already taken a brief look at these bonobos in the context of sexual behaviour amongst immatures. That was a study from the early 1980s when these bonobos were not yet habituated and, as they have never been provisioned, these early observations under natural conditions were more difficult than those at Wamba. Nevertheless, in 414 hours of observations only 69 copulations between sexually mature males and females were seen (Thompson-Handler, Malenky, and Badrian 1984).

Foraging parties at Lomako tend to be small, often with only 2 to 5 members (excluding dependent offspring). Over at Wamba, most bonobo sexual activity occurred in larger parties so we at least get a general picture that smaller foraging parties are going to be a bit uneventful as far as sexual activity goes. It is, in fact, possible to follow such foraging parties for days and not see any copulatory sex at all.

We do have some numbers from a later study of the Eyengo community at Lomako, covering 196 days during the period 1993 to 1995 (Gerloff et al. 1999). In this study the copulation rates for the 8 sexually mature males (with the 18 sexually mature females) ranged from 0.22 to 1.29 per observation day. The highest ranking male had the highest copulation rate which should no longer come as any surprise.

It is also interesting to note the sex ratio of this community: more than twice as many females as males. Bonobos are often presumed to differ from chimpanzees in having a more equal sex ratio, often attributed to the lower level of violence (and therefore mortality rate) amongst the males of the more peaceful species. But bonobo communities beyond the E1 community at Wamba (with its equal sex ratio), such as this one at Lomako, contain a lot more females than males, not unlike many chimpanzee communities. Rather than female-biased sex ratios in chimpanzees contrasting with equal sex ratios in bonobos, we also find female-biased sex ratios in many bonobo communities, and the two species do not differ in this respect after all. This means that a female-biased sex ratio cannot be used as evidence of a high level of male violence amongst chimpanzees, unless the same reasoning is also applied to bonobos.

Nevertheless, whatever the reason for variation in sex ratio, the 18 available females in the Eyengo community makes the copulation rates for the 8 males seem particularly low.

Though studies of captive bonobos are of limited value, it is useful to note that even in these conditions bonobos can show little heterosexual activity. In a study carried out at Planckendael Zoo in Belgium, copulations were found to be "rare" (Vervaecke *et al.* 2003). This captive group's three sexually mature females and four sexually mature males were studied from January to March 1995 and from November 1996 to February 1997, making a total of 5.5 months of continuous daily observations. The three females mated only 26, 13, and 8 times over this whole period; far lower figures than we would expect, especially in a captive group.

The females also showed preferences for particular males and did not copulate with all of them, whereas they were not at all choosy when it came to the much more frequent non-copulatory mounts (there were 102 of those). This difference between non-copulatory sociosexual mounts and actual copulation shows how these 'sexual' behaviours cannot all be lumped together as one and the same. The bonobos seem to know the difference even if humans often find it difficult.

We cannot find a simple answer to our "how much sex does a bonobo get" question but we have at least gained some insight into individual variation and the circumstances where sex most often occurs. Very occasionally, certain individuals may have sex multiple times in one day but, overall, the picture is much less orgiastic. Some, especially the higher-ranking males, may be having sex more than once a day on average but for other males, they would be highly envious of "Alvy Singer's" three times a week.

A further observation made by Kano also needs to be considered. Kano writes that he does not think bonobo males are always ejaculating during their copulations. He says that at Gombe, Jane Goodall reported that ejaculate could often be seen adhering to the chimpanzee's penis. In contrast, most of the time at Wamba there was no evidence of ejaculation, and the penis was frequently still erect after copulation. So it is quite likely that many of these bonobo copulations are more like brief, genital-contact rituals, in line with their social rather than sexual nature.

Researchers at Lomako have also said that males often still had erections after copulation, and it was difficult to tell if ejaculation had occurred (Thompson-Handler, Malenky, and Badrian 1984). Copulations at Lomako lasted 12.2 seconds on average but ranged from 1.5 seconds to 45 seconds; the 'blink and you miss them' copulations at the lower end of this range are more likely to be brief, non-ejaculatory copulatory mounts.

Back at Wamba, Kitamura writes how females at the feeding site presented to males who had food, repeatedly begging and following the male. Though the male would often mount such a female, this did not always result in complete intromission and ejaculation (Kitamura 1989). This again points to a distinction between sex proper and the more formulaic sociosexual interactions.

Bonobo copulation is usually silent but sometimes the female screams at the end. While we might presume that this signals a sexual climax in the female, and ejaculation by the male, in one case seen by Kano a female screamed while copulating with a male that had no external genitalia except for a stunted penis. Not a pretty image, granted, but it points to how 'faking it' might be involved in ritualised social sex, making it even more difficult to distinguish a social ritual from sex proper.

Other reported instances lend further support to the existence of this ritualistic nature of sexual encounters. One such incident was when a mature male pushed his buttocks in front of an adult female and she responded by standing behind him in a mounting-like posture. The female then grinned and screamed though there was no genital stimulation whatsoever. In another incident, a mature female mounted another mature female, then she pressed the genitals of the mounted female and the mounting female (not the mounted) screamed. Human observers assume that grinning and screaming is indicating orgasmic pleasure but this is evidently not necessarily the case.

Even infants involved in pseudocopulatory interactions may scream and grin as the mature bonobos do, so we need to take into account the learned social signalling nature of these behaviours, and that they are not necessarily indications of sexual climax, even in adults. It is not that real sexual pleasure is completely absent from all of these encounters, only that there is a ritualistic, social convention nature to many of them. In this way, they often stop well short of being sexual encounters that have been provoked by a straightforward lustful desire in the way the human observer often imagines.

Surely, though, the frequency of bonobo heterosexual sex outstrips that of the chimpanzee, doesn't it?

Kano points out how difficult it is to compare studies of the two species because the numbers of individuals, the ratio of the sexes, and the context of the studies varies across communities of both species. We have seen how the artificial feeding sites affected results, and how the rate of copulation for the same Wamba bonobos was much higher under provisioning than in wild conditions, so even the behaviour of the very same bonobo community differs according to context (Kano 1992).

Early reports from captive bonobos told us that females mate throughout their sexual cycles, placing bonobos alongside humans as the two species where sex and reproduction have been separated. From this we would expect the mating frequencies of the chimpanzee and the bonobo to be very different but no, studies in the wild have shown them to be much the same.

A comparison of Gombe and Mahale chimpanzees with Wamba bonobos found the copulation rates of adult male bonobos at Wamba to be equal to or lower than those of the adult male chimpanzees. This is not what we would have expected. As for the adolescent males, the young chimpanzees were getting more sex than their bonobo counterparts (Takahata *et al.* 1996).

This study by Takahata also noted a difference between the adolescent females of the two species. Whereas adolescent chimpanzee females (at least in these populations) rarely mated with adult males, the bonobo adolescent females had a high copulatory rate with their adult males, and higher than that of the adult females.

Sexually mature males are constantly fertile and potentially interested in sex but things are more complicated when it comes to the females. This is because of the changes that occur during the female's sexual cycle, and changes that occur throughout the female's lifetime depending on her maternal status.

For cycling females (and the occurrence of sexual cycles is important to note here) the copulation rate, if averaged out over the whole cycle period, was found to be about the same for the females of the two species. But it is where the copulations occur within the sexual cycle that is the difference: chimpanzee females had a much more concentrated mating period coinciding with their maximum swelling phase, whereas the copulations of bonobo females were spread out over more days of the sexual cycle.

As well as this difference between the females of the two species within their sexual cycles, female bonobos also have many more of these cycles throughout their lifetimes. There is much individual variation in both species but overall, bonobo females have more sexual cycles with swellings. This is because they often resume their (initially non-fertile) sexual cycles within 15 months of giving birth whereas chimpanzee females can take at least twice as long to resume theirs. It seems counter-intuitive then, that this extended sexuality of the female bonobo does not lead to clear reports of far more copulatory sex going on in bonobo society.

It might be the case that the above study groups of chimpanzees at Gombe and Mahale have given us higher male copulation rates than generally occur across chimpanzee populations, possibly because of the particular ratios of males to females or the particular circumstances of these studies. Perhaps. But a later study of a new community of chimpanzees in the Kalunzi forest in Uganda found the copulation rate of these chimpanzee males to also be about the same as that of bonobo males (Furuichi and Hashimoto 2002).

It could be that the particular figures for the Wamba male bonobos were unusually low for some reason, but we have seen that in other bonobo communities, even when there are many more females than males, the copulation rate is far lower than we have been led to expect.

Whatever the specifics of these particular communities of chimpanzees and bonobos, these studies show us that there is not a clear and obvious distinction between the two species. We tend to believe that male bonobos are getting much more sex than chimpanzees, and that this goes a long way in explaining why they are more peaceful. Because such a distinction between the two species has not been found, it is unlikely to be a higher frequency of copulatory sex that leads to more peaceable behaviour in the bonobo male.

Perhaps the bonobos spread it about more equally?

Not according to the Takahata study which found that bonobos do not copulate with a greater number of different partners than do chimpanzees. It appears that bonobo females are not spreading themselves about more, and neither are a few chimpanzee males able to completely monopolise their females. Takahata's study also found no clear correlation between male chimpanzee copulation rates and rank, although the alpha male chimpanzee tended to be possessive near the time of ovulation.

We'll come back to male sexual competition later but for now we'll just note that a clear difference between the two species in their frequency of heterosexual sex for adult males does not seem to exist. And if we were able to compare the numbers of actual ejaculatory sexual encounters of bonobos and chimpanzees, it is even more doubtful that bonobos would come out on top.

A distinction we still have, though, is that female bonobos have sexual swellings well beyond the limited periods of fertility. Because bonobo females have their attractive sexual swellings for so much more of their lifetime it is assumed that there must be a lot more sex going on.

The general and simplified picture of chimpanzee females is that they have a few months of fertile cycles when they associate with males. During each of these cycles (lasting about 35 days) they have about 10 days when they sport a large swelling and mate a lot with males who compete aggressively over them. The swelling then disappears until the next fertile period. The female conceives within about 6 months, and when she gives birth she stays away from the males, foraging alone or with other similar females. When her youngster is 3 to 4 years old and is being weaned, she will again start having sexual cycles and will seek out the males.

In contrast, the bonobo has been said to have more or less continuous swellings throughout all this time, she is said to have sex throughout all this time, and the males are said to have so much sex available to them that they have no need to compete with each other for this vital resource. The continuous swellings of the females are also said to make ovulation impossible to detect so it is pointless for males to even bother trying to monopolise a female when they can have no idea if or when the precious egg is released.

The belief that female bonobos have more or less continuous sexual swellings and that they are more or less continuously sexually receptive comes from early captive studies. But these captive females were often still young, and we now know that the almost continuous swellings and continuous sexual receptivity is a feature of adolescent females. As soon as information came in from bonobos in the wild it was obvious that things were different, and we had the reports from Kano at Wamba that the frequency of sex for the female bonobo varied depending on her age and her reproductive state (Kano 1992). And yet, the belief that bonobo females are continuously sexually attractive and sexually active remains.

It is true that bonobo swellings don't disappear in the way chimpanzee swellings do. For chimpanzees there is a relatively straightforward cyclic nature of swellings, progressing from nothing to large, firm swellings around the time of ovulation. Bonobo swellings do not follow such an obvious cyclic change, and they can stay relatively noticeable throughout the cycle though they do become very soft and wrinkly. In bonobo females it is the firmness rather than the size of the swelling that follows a cyclic pattern, and it is maximum firmness that signals fertility in the way size does for the chimpanzee (Furuichi 1987).

In a study of the E1 community at Wamba, Furuichi found that 6 of the 10 females *only* mated when their swellings were at maximum firmness. Over 80% of all copulations occurred when the female's swelling was at maximum firmness, and the percentage jumps to 95% when slightly wrinkled swellings are included.

What's more, a bonobo female with a maximum swelling, unlike a chimpanzee female, will not necessarily be sexually active. Furuichi found that pregnant females and the mothers with young infants had maximum swellings but seldom copulated. Also, one adolescent female had a long period when she was maximally swollen but she only copulated in restricted periods within that time span.

As well as the adolescent females having small but visible swellings throughout the sexual cycle, Furuichi reported that there was also a lot of individual variation amongst adult females in the maximum swelling period, with the older females having maximum swelling phases lasting from 3 to 22 days, which is quite a range, in cycles lasting from 37 to 49 days.

In a later study of the E1 community, which looked at copulation attempts by males, Furuichi and Hashimoto (2004) found that 11 of 32 copulation attempts occurred when females did not have their maximum swellings, and the females accepted 5 of these attempts. So male bonobos do prefer females when they have their maximum swellings but they do not restrict their sexual interest in females to this phase.

When chimpanzee females have their maximum swellings they are the focus of a lot of male attention and will mate at least once every two hours, which is a much higher frequency than female bonobos with their maximum swellings. But female bonobos have swellings that don't completely disappear, and they can still attract some sexual interest from the males beyond the period of maximum firmness. So the pattern of female sexual activity between the two species is different. Chimpanzee females have more concentrated periods of sexual activity whereas bonobo females have sex over more extended periods but they end up having much the same amount of sex per oestrous cycle; it is just spread out differently over the cycle.

What we don't have, though, are adult bonobo females with constant swellings that are constantly sexually attractive to males. The early image of the continuously sexually active bonobo female is yet another one that drops away when our knowledge expands beyond adolescents in zoos.

There is one consequence of the reduced connection between sex and actual female fertility in bonobos that the wannabe bonobo is unlikely to welcome: female bonobos were found to show a less positive behavioural response towards copulation than female chimpanzees. In other words, female bonobos were less actively interested in sex than were female chimpanzees. This finding has also been confirmed in the later study of the Kalunzi forest chimpanzees, as the females there were also found to actively initiate sex more often than female bonobos (Furuichi and Hashimoto 2002).

Chimpanzee copulation occurs almost exclusively in the maximum swelling phase of the cycle, and female chimpanzees are usually keen to engage in these fertile sexual interactions. This is the sexually rampant female behaviour seen in many species when females are in oestrus. Hanging around with the males is not something female chimpanzees might want to do for longer than necessary if it means more competition for food, as well as living with the less pleasant side of male chimpanzee behaviour. When a female chimpanzee is in oestrus she is going to be the focus of a lot of male sexual interest which can become very stressful for her; she wants to get the job done and return to a quieter life away from the males.

There is not a total disconnection between the sexual attractiveness of female bonobos and their actual fertility but there is a great dilution. This means that when females do have maximum swellings they are not going to be inundated by sexually voracious males ready to pounce. Bonobo females can hang around with males pretty much all the time; access to food is not a problem for them when traveling in mixed-sex parties, and male behaviour in general is not so much of a problem either, at least compared to chimpanzees.

With the extension of the bonobo's sexual activity beyond her fertile phase, she does not become the focus of male sexual attention in the way her chimpanzee sister does. On the rare occasions when female bonobos are actually ovulating they are still going to be physiologically motivated to mate in order to conceive but, with ovulation a much less obvious event than in chimpanzees, they are also going to have more control over the situation. Female bonobos can be particularly choosy about their sexual partners when they are actually fertile.

Beyond the occasional periods of actual fertility the female bonobo's motivation to mate will depend more on external factors than internal. Because she is not experiencing the sexual urges that come with true oestrus, her sexual activity at these times is something that is going to be a response to male behaviour and to particular situations rather than it being the proactive, female urge to mate that is tied more to her fertile period. The bonobo female is never going to be as wildly libidinous as her fertile chimpanzee sister.

At this point, though, we have to note that the differences between chimpanzee and bonobo are, once again, not as clear-cut as we might have liked to see. Furuichi and Hashimoto (2002) also note in their study that the chimpanzees of the Taï Forest do not follow the standard chimpanzee picture of little association between the sexes beyond the rare occasions when adult females are actually fertile.

Taï chimpanzees also show frequent association between the sexes, and there are more mixed-sex parties in these chimpanzees than are normally found in other known chimpanzee populations. Interestingly, more than half of the Taï females have been found to resume sexual swelling cycles within a year of giving birth rather than the two-and-a-half years or more that is normally associated with chimpanzees. Unless these Taï chimpanzees think they are bonobos, we are again seeing the overlap between the two sister species rather than the oft-stated chimpanzee-bonobo polarisation.

In the way that so much of what we have learned about bonobos has come from Wamba, and especially the artificial feeding sites, much of our picture of chimpanzees comes from Gombe and other similar eastern populations such as the Mahale communities. But in the west of Africa there are communities in the Taï forest that have shown behaviours more similar to those of bonobos. We will look more at these Taï chimpanzees later but for now just note how information from more populations of the two species has increased our awareness of the variation and overlap amongst chimpanzees and bonobos.

Political unrest in the Congo prevented much research during the 1990s, and many researchers evacuated their study sites in 1991. There was civil war in 1996, war again in 1998, and finally a ceasefire in 2002. Research resumed in Wamba in 2003. Since then, there has been no artificial provisioning of the Wamba bonobos so it is useful to look at a more recent study from there, based on over 1000 hours of observations over 134 days during the period from September 2011 to December 2012 (Ryu *et al.* 2014).

In total, 208 copulations were observed involving 9 adult females with 7 adult males and 3 adolescent males, so that's only one every five hours amongst all these bonobos. Females with maximum swellings were involved in most of this copulatory sex, so the frequency of copulation was clearly related to swelling status. Young females who had been in the community for less than 5 years were having much more sex than other females, even outside of their periods of maximum swelling, while two old females never copulated with males outside of their maximum swelling phases.

These results confirm earlier findings, and the authors conclude that the greater sexual activity of younger females indicates that they have a greater need for sociosexual behaviour. They suggest this is due to their vulnerability as immigrants in competition with established community members. These young females, who are not going to be fertile and conceive for a number of years, may be using sex as a way of gaining support from males, or they are using sex to appease males and reduce potential aggression from them.

This fact of young, vulnerable females engaging in a lot more copulatory sex than established females shows us, on the one hand, how sex is a useful female strategy to avoid or reduce potential social conflict. On the other hand, it raises the question of why those older, more dominant females are not indulging so much in what is meant to be a mutually pleasurable activity. Female power amongst the bonobo seems to be about having a lot less copulatory sex, and female extended sexuality is more of a social tool to be used when needed rather than a simple extension of her lust for sexual pleasure.

Since the early captive studies of bonobos, which told us that female bonobos have almost constant sexual swellings and are almost constantly sexually receptive, the picture has been greatly revised. That early portrayal was based on young female bonobos but we now know that as these females mature, and as they become mothers, their overall frequency of sexual activity becomes something not so obviously different from the chimpanzee. Not surprisingly, more recent captive studies also contrast with the early ones simply because those young bonobos of the early studies have now grown up.

To sum up the picture so far, when chimpanzee females have their large sexual swellings there will be a lot of sex involving those females. These are much more likely to be potentially fertile matings than those of bonobos, and the purpose of chimpanzee sex is to conceive the next offspring. If we just consider females when they have maximum sexual swellings, chimpanzee females have a much higher frequency of matings than do their bonobo sisters.

Bonobo females have a lower frequency of copulatory sex during their maximum swelling phase but they do copulate more often outside of this phase of their cycle. Bonobos also have more swelling cycles during the interbirth period (many of them non-fertile) but their frequency of matings is relatively low for much of this time. Adolescent females of both species have extended periods of swellings and are keen to mate but chimpanzee adolescent females are not as attractive to adult males as are their bonobo counterparts. It is low-status, adolescent females that are the stars of sexual activity amongst bonobos.

We'll look now at one particular feature of bonobo sexual behaviour that has often grabbed the headlines: face-to-face mating. The figures from Frans de Waal's San Diego Zoo study for heterosexual mounts were: 162 face-to-face mounts and 33 "doggy-style". Other captive studies had also shown a high proportion of face-to-face sex so this was believed to be the norm for bonobos. It's not.

Initial studies at Wamba reported that fewer than 30% of copulations were in this position, and these mostly involved adolescents. One of Kano's fellow researchers at Wamba carried out a study in the late 1970s and found that 233 (90%) of the 258 matings were doggy-style. In Furuichi's study of the E1 community in the mid-1980s, 94 (86%) of the 109 copulations were doggy-style. So the face-to-face position turns out to be far less frequent in the wild, and it is particularly rare amongst the adults.

Similar results came in from Lomako where face-to-face mating was found not to be the normal adult posture in the wild, only occurring when younger males were copulating (Thompson-Handler, Malenky, and Badrian 1984, White 1992).

Even in captive groups, as the young bonobos matured their sexual position changed. In the late 1970s it had been reported that captive bonobos at Yerkes preferred the face-to-face position. Later studies at Yerkes found almost all mating was now doggy-style – the male was now an adult (Blount 1990).

It has become evident that the high frequency of face-to-face matings reported in captive groups of bonobos has been mainly due to the young age of these captives. Kano writes that when a female at Wamba presented for a face-to-face mating the male would not want to mount but this kind of refusal did not happen when the female presented him with her rear. Occasionally, though, a female would move from this position to the face-to-face position after copulation began.

Face-to-face mating does occur in bonobos but rarely so when adult males are involved. Females do appear to prefer that position, perhaps because of the more frontal positioning of their genitalia, but they know what the adult male prefers and they mostly oblige.

And finally, for those wannabes who picture multiple individuals involved in the same sex bout, this turns out to be a very rare occurrence amongst mature bonobos. As with so much of the 'anything goes' bonobo sex, it is again the juvenile males who are the ones looking to join in with others. Sometimes an adolescent male may continue with this juvenile behaviour but it has disappeared in the adults (Kano 1992).

So, contrary to what we may have imagined, threesomes (or foursomes or more) are not a feature of sexual behaviour amongst sexually mature bonobos, and another 'Kama Sutra' image is gone.

So far in this chapter we have discovered that the frequency of heterosexual sex is not so different between chimpanzees and bonobos, and face-to-face mating is not the norm for bonobos after all. Female sexual swellings are different between the two species, and female bonobos are sexually attractive and sexually receptive for more of their lifetime than are chimpanzee females, but chimpanzee females are much more sexually proactive. Overall, the frequency of sex for the males of the two species turns out to be much the same. And it is the low status adolescent females who account for most of the bonobo copulatory sex.

Bonobos are more often in mixed-sex foraging parties than are chimpanzees but at least one known chimpanzee population, that of the Taï Forest, has a similar degree of association between the two sexes. It is likewise notable that many Taï females also resume sexual cycling within a year of giving birth. There is a definite overlap between bonobos and chimpanzees, and we cannot easily say from anything we have looked at so far that there are clear distinctions between the two species that give us an obvious answer as to why one is more peaceful than the other.

We have already touched upon the variation between male bonobos in their frequency of sex with females, and it has been mentioned how male rank plays its part in this variation, so we'll end this chapter by looking at male sexual competition. The image we have of bonobos is one where there is no sexual competition, where bonobos are egalitarian and indifferent to the sexual activity of other members, and where bonobo life is basically one big love-in. We have already come across powerful pointers that this is not so, such as the chasing of lower-ranking males from the feeding site and thus from the females, so now we'll see what other evidence exists.

First we return to Kano's early studies of the bonobos at the artificial feeding site, and see what else he tells us about male sexual competition. Of 515 copulations recorded at the site, another bonobo interfered in the copulation 33 times. Adult females were responsible for 6 of these interventions, and adult males for 27. In 8 of the 33 incidents it was simply a matter of threatening and pursuing the target, while on 10 occasions the victim received an open-handed slap. These behaviours, though few in number, are not what we would expect from the bonobo. Even if we can excuse these 33 incidents as relatively infrequent, it turns out that there is more going on here than meets the eye.

Kano describes how a male often solicited a female with an erect penis display. If there was no response from the female he approached her and gently touched her head, shoulder, back, or knee. The male would retreat, display, and be approached by the female several times, and in this way he would draw the female away from others before he copulated with her. Males also intentionally solicited females who were already separated from others, precisely to avoid any interference. So, rather than an indifference to the sexual activity of their fellow bonobos, males were found to be acting in ways to hide what they were doing, and therefore they avoided provoking any negative reactions from others.

Kano (1996) carried out a further study at Wamba to specifically look at male sexual competition, noting how this is presumed to be lower in bonobos because more oestrous (or pseudo-oestrous) females are available to males. When he looked at the rank of males responsible for copulations across subparties, he found that about half of copulations involved the top ranking male in the party. Dominance rank of male bonobos was found to affect their chances of mating, and male-male sexual competition is, Kano concludes, more intense than previously thought. The effect of dominance was particularly strong in small parties where only the top-ranking male had clear priority in mating.

In Kano's earlier study (above) he had found a low rate of aggressive interference by males (only 27 cases in 515 copulations) but males were also found to be hiding their sexual activity. With the new study results showing the effect of male dominance on copulation frequency, he had found further evidence for sexual competition. Open aggression is fairly rare, Kano concludes, because males choose to refrain from initiating sex when this may provoke another's aggression or harassment.

The absence of serious open aggression between males in competition over females is not because they are all equally involved in mating with females, and it is not because males are indifferent to the mating activities of each other; it is because the lower-ranking males often don't even try to mate unless a higher-ranking male is not around. Kano further suggests that taking first-rank in a subparty is not so different from the chimpanzee consortship where a male and female pair travel for a period of time away from others. We could even view the subparty as a temporary version of a gorilla group: a number of females with one or more males, and the top-ranking, dominant male doing most of the mating with the females.

One other factor that can help a male is the presence of his mother. Kano found that three adult males with mothers still alive were more dominant and had higher mating frequencies than three adult males with no mother. He also found that younger males with mothers were able to out-rank motherless males in their prime. Having a mother still alive can raise the rank of a male, and enable more access to the females.

What information was coming in from Lomako?

At Lomako, when the males arrived first at a big fruiting tree the dominant male would try to evict all other males from the tree before the females arrived. He then occupied the main access route to the tree, and when the females appeared they had to mate with the male before they could feed in the tree. In other situations, when a male joined a foraging party he would be looking to get close to the females and to monopolise them, though this also required female cooperation. If more males joined a party the effort to monopolise females became too much, and it was no longer possible (White 1992, White and Lanjouw 1992).

Also at Lomako, both sexes were observed to have preferences for certain mates, and females were observed to become choosier around ovulation, sometimes going off with just one male at this time. And finally, though rare, mating was seen to occur in response to male aggression, i.e., sexual coercion is not, in fact, totally absent. These findings from Lomako are facets of the forgotten ape that rarely get a mention.

While information coming from the two field sites, and even from the same bonobo community, can vary, there is clear evidence of sexual competition and that bonobo life is not as rosily egalitarian as many wannabes imagine.

So far we have covered the published research up to the time of Frans de Waal's 1997 book, *Bonobo: The Forgotten Ape*. In his book, de Waal incorporates much of the newest information at that time, writing that sexual competition in bonobos is more intense than previously thought, and it is based more on fear of what might happen rather than actual fights. Bonobos are reported to be timid and alert, and quite fearful. This puts something of a different slant on interpretations of bonobo society: if it is the laid-back, nothing-to-worry-about, everyone is equal and all get along just fine society, then why so much emotional agitation, timidity, and fear?

Low-ranking males, de Waal writes, learn to be secretive about their sexual exploits, and they develop all sorts of sneaky tactics to attract females, such as dropping twigs on to a female from above. Is this what we would expect from an egalitarian species where sex is openly shared? Sexual competition is just more subtle, not absent.

What does make bonobos different is, as de Waal says, the involvement of mothers in male status. Why should the status of a son matter to his mother? It is because mothers enhance their reproductive success through sons. When a mother improves the reproductive success of her son she is leaving more grand-offspring, and therefore she has improved her own reproductive fitness. Male status matters to males and to their mothers; as de Waal puts it, "females have little to compete over except their sons' careers".

We will be looking more at females later, and at the sexual behaviour of bonobos beyond heterosexual copulation, but first we need to see if we can discover anything further on male sexual competition in what has been published since 1997.

DNA testing of the Eyengo community at Lomako (Gerloff *et al.* 1999) found that 5, and possibly as many as 7 of the 10 offspring were fathered by the two highest ranking males who were the sons of two old high-ranking females. Although overt aggression was rare, the two top-ranking males did displace each other as well as lower ranking males in order to monopolise access to females. There would appear to be a non-violent acceptance of the status and mating rights of higher-ranking males. Males who were in conflict with other males also received support from their mothers. Acting as her son's 'wingmom', a bonobo mother can improve his status and paternity success, and she gets herself more grandkids.

In 2002, a new bonobo research camp was established at LuiKotale. These bonobos were habituated without the use of provisioning, and a study began in mid-2007 specifically looking at the influence of mothers (Surbeck, Mundry, and Hohmann 2010). This community numbered up to 35 members, including 5 adult males, 4 adolescent males, 11 adult females, and up to 5 adolescent females. Note again the sex ratio, with more than twice as many adult females as adult males rather than the equal sex ratio that was thought to exist in bonobo communities.

The male bonobos at LuiKotale formed a strong linear dominance hierarchy, which would be unnecessary if male rank did not matter to bonobos; it clearly does. Of the 9 sexually mature males, 6 had their mothers in the community, and these males were traveling with their mothers for 81-92% of the time. This study found that the presence or absence of a male's mother did not in itself predict the dominance status of a male, but when lower-ranking males had their mothers around, their mating success improved. When there were no mothers in a foraging party, the highest ranking male in that party had over 40% of all the copulations with females with maximum swellings, but when all the mothers were in the party, this fell to 25%. Overall, high ranking males had more sex with the most attractive females but the middle- and lower-ranking males improved their success if their wingmom was around.

This study also recorded 134 aggressive interactions during conflicts over access to oestrous females. Of these, 95 involved two males, 37 involved a male and a female, and 2 involved two females. In 6 of the 37 male-female cases, the aggressive interaction involved the oestrous female, while in 30 cases it was the mother of the male who was trying to mate who had become involved. In 13 of these interventions by a mother, she either intervened to stop an unrelated male from mating so that her son could, or she supported her son when an unrelated male tried to interfere in his mating.

The presence of a mother, therefore, was not found to be essential for a male to achieve high rank but mothers did improve the mating success of low- and mid-ranking sons. When mothers were absent from foraging parties, dominant males were more easily able to restrict the mating activity of subordinates.

The rate of actual physical intervention by mothers though, was fairly low, so mothers were also improving their sons' mating success in a different way. Because females forage together, the presence of mothers meant sons could get closer to the oestrous females in the party; a male traveling with his mother is also getting close to the objects of his desire.

Finally, this study also found that mating efforts by males were clearly more pronounced when females had their maximum swellings, confirming again that female bonobos are not constantly attractive to males. There was no evidence of male aggression against females in the context of mating, but neither did females try to avoid the mating efforts of dominant males. It looks like females either found dominant males irresistibly sexy or they felt it was better not to refuse such males.

How much sex does a bonobo get? Far less than the wannabe bonobo imagines. Extended female sexuality is not as extended as we used to think, and the maximum sexual swelling is clearly what makes a female most attractive to males. The ever-willing adolescent females, with their smaller but firm and more constant swellings, are the females pumping up the mating frequency; these females use sex to gain acceptance in their new community, and to gain access to food. The adult males, meanwhile, are only really interested in doggy-style sex, and males exist in a clear linear dominance hierarchy where rank matters when it comes to getting that sex. Mothers also influence the mating success of males, either with actual physical intervention or by enabling proximity to oestrous females.

The relationship between mother and son is an important one in bonoboland. Daughters are making plans to leave from the age of 7 or 8, initially becoming peripheral to the community and then settling for good in a new community by about the age of 10. But for a son, he will have a close and important relationship with his mother for as long as she is alive. This raises an interesting question about bonobo society. If alliances between females are important in keeping male behaviour in check, and mothers also support sons, what does a mother do if conflict occurs between her female ally and her son?

The single study we have that specifically tackled this question was carried out on a captive population at Planckendael Zoo in Belgium. In this study, the female supported her coalition partner over her son because this particular female's social status in this zoo group was more dependent on supporting the higher-ranking female than her own son (Legrain *et al* 2011).

We cannot take too much from this single result from a small captive population, and the question of where a mother's allegiance lies is one that still needs to be examined. Other studies of captive groups have shown that female alliances weaken when the females become mothers. When conflicts occurred between females and the offspring of their former allies, mothers were rarely inclined to support their former friends. Instead, they would withdraw and make appeasement gestures to both parties in the conflict (Stevens *et al.* 2006, Stevens *et al.* 2008).

Do we have strongly bonded females cooperating to collectively dominate males? Or do we have mothers acting for the benefit of their sons? The reproductive fitness of a bonobo mother and son are tied together, and this creates an interesting convergence of male and female interests that could be at odds with the potential benefits of female allegiance. Bonobo females who are mothers of adult or sub-adult sons are likely, at times, to be facing some interesting dilemmas.

We will look at examples of mothers supporting sons when we look at aggression later but next it is time to turn to bonobo homosexual behaviour, and how this does, or does not, show bonding between members of the same sex.

Three: homosexual sex

If it isn't the frequency and distribution of heterosexual sex that has set the bonobo apart from the chimpanzee, what is it? A well-known aspect of the fabulously cool nature of bonobo sexual behaviour is their same-sex shenanigans. Are the big male bonobos making love, not war, with each other? And how does that play out, considering what we now know about the male hierarchy and how male rank affects mating opportunities? Male sexual competition plays a bigger part in bonobo life than we used to imagine, so where does that leave their male homosexuality?

Frans de Waal's San Diego group initially painted a picture of laid-back sexual encounters between the adult male and the two young adolescent males but we then discovered how sexual competition eventually led to the removal of the adolescents from the adult's group. Before tensions had risen to that level there had been 34 sexual encounters in which the adult male had massaged the presented adolescent penis following a conflict. As much as that sounds like a fun way for males to resolve their conflicts, this is not a behaviour that has been reported in the wild. What we do have in the wild are the other male-male behaviours observed in the zoo: mountings of males by males, rump contacts between two males, and, though rarely, face-to-face genital contacts, sometimes called penis-rubbing. It is time again to return to Wamba.

Kano's 330 hours of observations at the artificial feeding site in the late 1970s gives us figures of 103 mountings, 31 rump-rump rubbings, and 2 gg-rubbings amongst sexually mature males (Kano 1992). These 2 gg-rubbings occurred when two males, hanging from a branch, swayed their hips while rubbing their penises together.

Kano tells us that the sexual behaviours between sexually mature male bonobos often occur in the context of aggression. To illustrate this, Kano describes how the attacking male bonobo "may spring on a male, who, cut off from escape, is grovelling and screaming, and the attacker will mount or rump-rub the victim. Or the attacker may confront his victim, suddenly facing the victim's buttocks and demanding mounting or rump-rubbing."

While this may well be an avoidance of "war" it does not sound much like "making love".

Sometimes a male may simply display to another male, much as he would towards a female, and mounting or rump-rubbing may (or may not) occur. But most male-male sexual behaviour, like most adult sex, occurred during the frantic first ten minutes or so on arrival at the feeding site. Kano (1990) describes one such scene from February 9, 1978:

A low-ranking male was the first to enter the feeding site. This male was immediately followed by a middle-ranking male who hurried towards him, so the low-ranking male presented his buttocks, and they rump-rubbed for a few seconds. A high-ranking male, Ude, then entered the site and the low-ranking male ran off to a tree. Ude then rushed toward, mounted, and made vigorous thrusting motions over the middle-ranking male who crouched and screamed. This was followed by two "mutually agreed upon" bouts of rump-rubbing between these two males, lasting 27 seconds and 45 seconds.

Then the alpha male Kuma arrived, dragging a branch, and he rushed towards high-ranking Ude who screamed and jumped out of his way. Kuma threw away the branch and presented his buttocks, turning to look at Ude who, still screeching, rump-rubbed with the alpha male. Ude then proceeded to seek out the middle-ranking male he had rump-rubbed with earlier, attacking, chasing, and mounting him while his victim screamed throughout.

Is this the kind of sex between guys the wannabe bonobo imagines? When reading such descriptions it is hard not to be reminded of some of those pretty nasty prison stories we sometimes hear. While the male homosexual activity of bonobos can be "mutually agreed upon", there is also a lot of crouching and screaming where there has been no such mutual agreement. Kano says that while he has never seen forced heterosexual mating, forced mounting occurs frequently between males. The male wannabe bonobo might want to watch out for that one.

The five-month study by Kitamura from the late 1970s recorded 39 mounts, 10 rump-rubbings, and 1 gg-rubbing (Kitamura 1989). Kitamura also writes that these sexual contacts amongst males are often preceded by aggressive chases but he noted that dominant/subordinate roles are not always clear: though the alpha male was always the mounter, other males sometimes switched between mounter and mountee roles.

Sometimes, Kitamura says, the dominant males presented in a threatening way, and we have seen examples of this from Kano above. He also makes the point, and it is an important one for the naked bonobo, that mounting does not involve penis insertion, and sometimes does not even involve thrusting. How much of a relief or a disappointment this is to the male wannabe will, no doubt, depend on his personal preferences.

Kitamura gives an example from his study which could be useful as we try to think how we might incorporate this crucial male homosexual bonobo behaviour into our own, wannabe lives. Yasu (an adult male) picked up a stick, ran with it towards Ibo (a young adult male), and assaulted him. Ibo presented to Yasu who mounted him. The pair separated and moved towards an area with sugarcane, but they then came close to each other again. Both males presented to each other, resulting in an incomplete rump-rump contact as they were at right angles to each other. Then Yasu mounted Ibo and, immediately afterwards, Yasu presented and Ibo mounted.

What a neat way to handle an assault and avoid a physical fight.

Previously, we looked at an early 1990s Wamba study by Hashimoto which was the main source of information on the sexual behaviour of immature bonobos. In that study, Hashimoto found that most adult male sexual behaviour was with other males rather than with females, and occurred in the context of conflict resolution. As with other studies at the Wamba feeding sites, most sexual behaviours occurred immediately on entering the feeding site, and about half of the male-male sexual contacts were clearly preceded by aggressive behaviour which was then ended by the genital contact (Hashimoto 1994, 1997).

What about the bonobos at Lomako? The Lomako study we looked at from the early 1980s, in contrast to Hashimoto's, saw hardly any cases of genital contact between males in the whole 18 months. The researchers at Lomako observed one mounting between adult males which, they say, resembled the dominance behaviour seen in chimpanzees and baboons. They also observed one episode of face-to-face penis-rubbing between an adult and an adolescent male, and one episode of rump-rubbing. In the latter case an adult male chased a smaller, past-prime male in a tree. The older male fled, squealing, but then suddenly stopped and presented his rear to the younger male who responded with a brief rubbing of his anus against that of the older male (Thompson-Handler, Malenky, and Badrian 1984).

The authors of this Lomako study note that dominance appeared to play a role in these male behaviours, though they only had the three incidents to go on. The only subsequent information we have from Lomako is that male homosexual behaviour continued to be relatively uncommon, and rump-rubbing and penis-rubbing were especially rare events (White 1992, Fruth and Hohmann 2006).

In one of these later Lomako studies, a single case of a mount with anal penetration was observed but no details of ages of participants (infants and juveniles are included in the figures) or context are given (Fruth and Hohmann 2006). Considering all the observations of males mounting males in captivity and at field sites, and that the male often has an erection, it is perhaps surprising such an incident does not occur more often. Researchers have consistently specified that intromission is not occurring during these mounting behaviours so this single observed incident shows that it is clearly not an intended goal, which is probably something of a relief to those cowering and screeching bonobos on the receiving end of forced mountings.

This is the sum total of information we have on sexual behaviour amongst adult male bonobos. What can the wannabe bonobo take from it?

Bonobos are not noted for close relationships between males, and though males sometimes simply solicit other males in the way they solicit females for sex, there is a clear and strong element of dominance behaviour running through most of these sexual encounters. This shows some overlap with the mounting behaviour seen between males in many species, and though the bonobo males also rump-rub and penis-rub, these are rare compared to mounting behaviours.

Much of this sexual behaviour between bonobo males is very brief contact, and it is often accompanied by a lot of anxious and fearful grimacing, squealing, screeching and screaming. While the genital contact does seem to calm the attacking males, and usually avoids an escalation into severe physical violence, it is difficult to really see human males turning the other cheek in quite this bonobo fashion. At the same time, it does not seem remotely close to the human form of male homosexuality – at least, not as it occurs outside prison walls, or, perhaps, some gay porn sites.

What about chimpanzees?

Frans de Waal (1989) writes about the chimpanzees at Arnhem Zoo in the Netherlands, and how two males would make up by kissing, mounting, and fondling each other's genitals. Male chimpanzees in the wild also often finger each other's scrotum (a gesture known amongst field-workers as ball-bouncing) at moments of mild tension, such as when parties meet and mingle. The mutual handling of genitalia by male chimpanzees seems to be an act of reassurance, and is presumed to be an indication of friendly intentions.

It is important to remember that chimpanzees have many ways to deal with the problems that will arise in any social group, and they live together quite peaceably most of the time. Chimpanzees will use an outstretched hand to beg for body contact, they will kiss to reconcile, and they will console another individual with an embrace. They will also share food after they have celebrated with loud vocalisations and body contact in which they hug and pat each other on the back. Chimpanzees are not antisocial brutes.

For bonobos, some form of mounting behaviour or genital contact is the predominant social behaviour that an overly agitated male seeks out, and this can immediately calm him and prevent his agitation turning to violence. In this way, potential conflicts can be quickly resolved.

Chimpanzee reconciliation can take longer to occur but they also have more varied behaviours, including the pant-grunt which they commonly use to acknowledge dominance relationships. The chimpanzee pant-grunt is a ritualised submissive gesture where the subordinate falls prostrate in front of the dominant while panting intensely. The dominant responds with a pacifying behaviour such as an extended hand or an embrace. Potential conflicts are thus avoided, and the submissive behaviour successfully serves to keep the peace.

This dominant/subordinate chimpanzee interaction may grate against our egalitarian leanings but is it really a better deal for bonobos? When bonobos are attacked they will grovel and shriek violently, and the attacker does not pacify his victim with a handshake or an embrace, but they will engage instead in a mounting or a rump-rub (Kano 1992). The fact that these behaviours derive from sexual behaviours does not make them necessarily pleasurable, at least not for both participants.

There are sexual elements in the social behaviour of many primates, whether as genital displays or mounting behaviours or genital inspections, including touching and licking genitalia. It is not unusual, therefore, for behaviours involving the genitals to be used socially, whether in greetings or to deter aggression or to reinforce rank-related interactions (Dixson 2012). The use of genitals need not be restricted to purely sexual functions, and when they are used in social interactions it is often not 'sex' as we would normally think of it.

Male bonobos predominantly use sexual or pseudosexual behaviour to deal with social conflict, but it is clearly not a required behaviour for the management of social relations. The non-sexual behaviours used by chimpanzees are just as effective for tolerance and bonding as the genital contacts used by bonobos (de Waal 1997). We might even say more effective, as bonobo males have very weak bonds compared to male chimpanzees.

Bonobos and chimpanzees both have effective ways of keeping the peace, so what are the differences between the males of the two species?

It does not seem from the actual figures that bonobo males are having more heterosexual sex than are chimpanzee males. Neither is sexual competition absent from bonobos, and they clearly have a male hierarchy and differential access to sex with females. As for sex between males, mounting behaviours occur in both species (and in many other species) though for bonobos it has a somewhat more egalitarian aspect to it when the role of mounter and mountee are reversed. We also have the occasions where both male bonobos present at the same time and they end up touching or pressing their rumps together. These contacts are not always due to mutual agreement, and the fact that a dominant male can force a subordinate to engage in what appears to be an egalitarian reconciliation shows that all is not necessarily what it seems in bonoboland.

The main difference between the males of the two species is in how they bond, or not, within a community. In both species, males stay in their natal community for their whole lifetimes yet it is only the chimpanzee males who form strong, if temporary, alliances with each other. Chimpanzee males compete for status within their community, and they will form alliances with other males to gain status but they will also put aside any within-group competition to work as a team to patrol and defend their joint territory, and sometimes to hunt monkey prey.

Behaviours between chimpanzee males are relatively easy for us to interpret: as members of our groups, whether families, social groups, or even nations, we can have serious internal quarrels yet drop these differences should any threat come from outside our own community.

In comparison, the behaviours of bonobo males can seem confused: males assert their dominance then act as if they made a mistake and don't really want to be the winner, even forcing their victim to engage in what otherwise looks like a mutually pleasurable sexual behaviour. There can be associations that occur between some bonobo males and not others but that's about as far as it goes for the male bonobo.

On the one hand, the absence of political alliances between males within the group, and the absence of bonding of all the males in defence of the whole community, means no joint attacks by male allies. On the other hand, without those very same reasons to bond there is also no motivation for joint anything. Bonobo males are not males that work together as a team; they are individualistic.

Blinded by their imaginings of unbridled sex, wannabe bonobos (or at least, male wannabe bonobos) have missed this essential aspect of bonobo behaviour: males have very weak bonds. For a male wannabe bonobo to truly release his inner bonobo in the pursuit of peace he needs to overcome that chimpanzee side of him that seeks bonds and alliances with other males. Our inner bonobo will have no inclination to be in, or to support, any male team; there are no bands of brothers amongst bonobos.

Our male ancestors, for better and for worse, did form alliances; how much these alliances were involved in inter-tribal violence is a matter of much debate but they were certainly involved in hunting. Once we have bonds between males we enter a whole different world from the one experienced by the bonobo.

Male bonobos have stronger relationships with females than with males, and they can have especially strong relationships with their mothers. It does not seem to be a greater frequency of heterosexual sex that has reduced their aggressive behaviour; neither does it seem to be their genital contacts with other males. What does matter is that they do not bond with other males.

It is now time to look at what is probably the most significant behaviour of this species: gg-rubbing between females. This is a frequent behaviour unique to the bonobo, and it is most likely the overriding reason for our imaginings of rampant and 'everyone with everyone' sex.

There is something in the heterosexual human male's sexual psyche that is completely beguiled by sexual activity between two females; while for some human females, such activity is about female solidarity and a rejection of men as mates. Consequently, this very same sexual behaviour is both a symbol of female power and independence from men, and a massive sexual turn-on for most of those rejected males. It is no wonder that the gg-rubbing female bonobo has such a following when she appeals to two very different audiences.

The sexual behaviour between female bonobos is quite different from that between the males: it occurs far more frequently, and it is not intermixed with open conflict and aggression. During the 330 hours of observation at the Wamba artificial feeding site in the late 1970s, Kano (1992) recorded 318 cases of gg-rubbing between sexually mature females, along with 5 cases of mounting between two females. This compares with 103 mountings, 31 rump-rubs, and 2 penis-rubbings between males.

The females who approached and begged for an invitation to gg-rub at the feeding site were younger, lower status females seeking proximity with the senior females. Adolescent, immigrant females generally kept a low profile but they would eagerly approach senior females for gg-rubbing, as well as readily responding to courtship displays by males. After engaging in these behaviours the young females could then carry their food and retreat to a safe place to feed in a tree or on the outskirts of the feeding site. These sexual behaviours with males or females are the young female's way of getting closer to the community's more dominant individuals, and are thus her way of getting closer to the food.

Kitamura's late 1970s study at the Wamba feeding site recorded 179 gg-rubbings, either in the trees or on the ground.

He describes how the two females move side-to-side in opposite directions, and how this needs to be synchronised or the rubbing may end immediately. Kitamura observed how females might present a number of times before one appeared to give in and take the upper position, and he saw a tendency for the younger females to take that upper position. This tendency for the lower-status female to take the upper position suggested that it was the least preferred position, and Kitamura concluded that females preferred to be in the lower position as this meant they were better able to make the side-to-side movements (Kitamura 1989).

A later study at Wamba (Hashimoto and Furuichi 1994) looked at 91 cases of gg-rubbing between females. Some of these cases were preceded by group excitement, including aggressive behaviour occurring between others in the party, and some cases followed sexual activity by one or both females. But for over 80% of cases, no triggering behaviour was observed. This means that gg-rubbing between females was unlike the sexual activity between males which often occurs in the context of conflict and aggression.

Early observations from Lomako recorded only 19 gg-rubbings between sexually mature females in 680 hours of observations. These gg-rubbings occurred most often during feeding, lasted from 5 to 34 seconds (average duration 14.82 seconds), and were generally silent; only twice did the female grin and make a nasalised vocalisation (Thompson-Handler, Malenky, and Badrian 1984).

Later Lomako studies also provide a somewhat different picture from that of the Wamba feeding site (White and Lanjouw 1992). In all the 23 visits by the Lomako bonobos to limited food patches, no gg-rubbings were observed. When parties arrived at a limited food source, the dominant females would simply exclude or displace subordinate females; not what we would expect of the bonobo sisterhood.

There were also no gg-rubbings in 24 of the 43 visits to superabundant food patches. As for the remaining 19 visits to superabundant food patches when gg-rubbing did occur (mostly at the beginning of feeding) this was not friendly bonding behaviour. These gg-rubbings were not 'good to see you' friendly greeting behaviours but were simply an expression of tolerance of another female's proximity. The researchers suggest that this could even be the signalling of an agreement to defend the food patch should such defence be necessary (because numbers matter for such defence).

At Lomako, gg-rubbings are described as rare events, occurring about once every six hours on average in a typical group of 5 or 6 adults. In fact, any sex in bonobo parties at Lomako is described as rare. And, rather than sex acting as a bonding activity, it was seen to be an activity that reduced friction between party members, especially in tense situations (White 1992).

This is an important distinction. The tension that bonobos feel when forced into close proximity with other bonobos is eased by the genital contacts but these do not create bonds, they only reduce the friction. This is about tolerance rather than a desire to connect.

Hohmann and Fruth (2000, Fruth and Hohmann 2006) carried out field studies at Lomako in the 1990s. The Eyengo community comprised about 10 sexually mature males, 20 sexually mature females, and 20 immatures. 1201 sexual encounters involving almost all 50 members of the community were recorded during this time, 661 (55%) of them homosexual, and 540 (45%) heterosexual.

Homosexual behaviour of males was found to be rare throughout all ages, while it increased from childhood onwards in females. Only 27 (4.2%) of the 661 homosexual interactions occurred between males, and most of these were mounts. The vast majority (95.8%) of homosexual behaviour was therefore between females, and most of this was in the context of feeding. When the two females were of different rank, most was initiated by the lower-ranking female, though there were 27 cases where the gg-rubbing appeared to be forced by the dominant female.

From the results of this study, the authors concluded that gg-rubbing has multiple purposes: it acts as a form of status acknowledgement, as tension regulation, and to some extent, as reconciliation. Because of its strong association with feeding and the tensions that arose around access to food, they did not find any support for the idea that gg-rubbing is a sign of affiliation or sisterhood.

The gg-rubbing was often occurring between females who were not particularly fond of each other, and most occurred between females of different rank, so the activity was not an expression of egalitarian bonds of friendship. And, because most of the gg-rubbing was limited to times of tension over food, it was not about females simply wanting to share a pleasurable encounter. Sex between females was occasionally used for reconciliation but mainly it was merely showing a tolerance of other females when feeding.

Back at Wamba, research resumed in 2003 following a decade or so of interruptions due to political unrest and war, and artificial provisioning was no longer used. In 134 days of observations during 2011 and 2012, 184 episodes of gg-rubbing were observed amongst the 9 females. Except for the old females, these occurred most frequently when females had their maximum swellings, and it was usually females with large swellings who solicited other females with large swellings (Ryu *et al.* 2014).

As in all the other studies, most gg-rubbing (90%) occurred in feeding contexts such as immediately after entering a feeding tree or after encountering a food item. Young females and others of low rank were using their swellings to ease the tension and get near to senior females who possessed the food.

Because females with large swellings were often soliciting other females with large swellings, it appears that sexual swellings are attractive to other females as well as to males. Though there were no adolescent females in this community at the time, the extended swellings of adolescent females may be as much to enable gg-rubbing with senior females as to attract the males.

If sexual swellings do make the females attractive to other females, this attraction is only acted upon during times of tension. Outside of feeding time gg-rubbing between females is only an occasional behaviour, so the females do not show much, if any, inclination to engage in this sexual activity simply for pleasure.

When we looked at the frequency of heterosexual sex in bonobos we found that male bonobos were not copulating more often than male chimpanzees, nor were they copulating with a greater number of partners (Takahata *et al.* 1996). Male bonobos do compete for sex with females, and they do have a hierarchy which correlates with their reproductive success. Though female bonobos are not as constantly sexually attractive and receptive as we used to think, they do have a larger proportion of their life than chimpanzee females when they have attractive sexual swellings, and they do spend more of their adult lives with other members of their community.

If extended female sexuality has not brought about a significant increase in the amount of sex available to bonobo males compared to chimpanzee males, the extended sexual swellings of females might have more to do with relations between females than between the sexes. Young immigrant females of both species have extended swellings that ease their acceptance into their new community. In chimpanzees it is about being attractive to the males, and the immigrant adolescent female chimpanzees will be traveling mostly with males when they first enter their new community. Immigrant bonobo females will be traveling with both sexes, and they need to get close to the dominant females in these foraging parties because dominant females will often have priority of access to food.

Bonobos use genital contacts at times of social tension, and immigrant females in particular are experiencing a lot of social stress. Kano (1992) writes that these young females seem to know well that they have no supporters or guardians in the group. He likens the situation to that of a new Japanese bride who bravely enters the groom's large household. Because the new bride is picked on by her in-laws, and her own family is far away, Kano says she is often under great stress and must be very patient.

Sometimes the young bonobo females can be attacked by resident females (Furuichi 1987), and whereas chimpanzee immigrants can look to the big males for support and protection, in bonobo society they have to ingratiate themselves with the dominant females too. Just as a sexual swelling distracts males from expressing aggression towards young females who want to take food from them, it seems that they can also distract dominant females.

For bonobo females, gg-rubbing is not something especially engaged in by females who like each other, nor is it some laid-back activity they indulge in just for pleasure. Engaging in gg-rubbing is more about relieving the tension between females who are not friends but competitors over food. It may even benefit dominant females by allowing enough females to come together to collectively defend a food patch. But they are not always willing to share that food, and, as the results from Lomako revealed, dominant females are often quite happy to simply displace subordinates.

Homosexual behaviour between male bonobos is relatively rare compared to that between females. Most of the male behaviour is in the context of open hostility and aggression; it is often an expression of dominance by one of the males, and it is often forced. Though this behaviour helps to reduce the likelihood of escalation to more serious physical aggression, it is not about bonding between males. Neither is it about egalitarianism.

Female status comes with age and number of offspring produced; you haven't made it as a female in bonoboland until you've had at least couple of kids. These established mothers have status and are at the core of bonobo society, and gg-rubbing mainly functions as a social lubricant to ease the tensions between females as they forage together. Young, immigrant females use gg-rubbing to get near to the senior females, and the almost continuous sexual swellings of these low status females help make this possible. Females do have close associations with certain other females, but gg-rubbing is not a particular feature of these friendships.

For bonobos, engaging the genitals does displace the tension and potential aggression that commonly occurs in social groups. But, even for bonobos, this does not always work. It is now time to take a closer look at bonobo aggression.

Four: aggression

How aggressive is the bonobo? If we still picture these sexy apes hugging and humping at the slightest hint of social conflict, we would expect the answer to be "not at all". But bonobos don't lack aggressive behaviours, and we met some of them when we looked at sociosexual behaviours between males. Competition and conflict are part and parcel of the life of pretty much every species, and bonobos are no exception. In this chapter we will look at aggression within bonobo communities, keeping in mind how little we have actually been able to see of bonobos deep in their natural, rainforest habitat.

We have already found how aggression, along with sex, occurs when bonobos come upon a food source. At Wamba, whether at the artificial feeding site or a naturally fruiting tree, the coming together of bonobos resulted in aggressive encounters and displacements of others from the food. Most of the aggression was between adult males, and at the artificial feeding site threats, chases, and attacks were so frequent that some party members were excluded altogether from feeding (Kano 1984).

In *The Last Ape,* Kano (1992) writes that one important element characterizing the relationship between adult males is aggressive interactions. At the artificial feeding site in the first two months of 1978, 163 aggressive or dominant behaviours were recorded involving two adult males, and another 14 were directed by adult males towards adolescent males. This compares to just 9 cases of aggression between two adult females. There were also 82 aggressive interactions involving an adult male against an adult female, 16 of an adolescent male against an adult female, and 11 cases of an adult male against an adolescent female. That's quite a lot of unfriendly behaviour for such a friendly ape.

The aggressive behaviour of bonobos abounds with variety, from physical contacts such as biting, hitting, kicking, slapping, grabbing, dragging, brushing aside, pinning down, and shoving aside, to glaring, bluff-charging, charging and chasing. In response to these acts of aggression, the submissive behaviours include flight, avoidance, prostration, grimacing, extending the hand, and touching the other's body.

When males approach each other it is not with relaxed warm feelings of friendly, pleasure-giving intentions but, as Kano says, the male's intention is one of dominance, and the prevailing mood is hostile. There is individual variation, as we would expect, and some males are very aggressive while others show little dominant behaviour. Some males will also have frequent social interactions with other males while others will rarely associate with another male in the same foraging party.

Kano found that males with mothers preferred to stay with their mothers rather than join other males, and he suggests that this is an obstacle to male-male bonding. Kano does not judge whether this "obstacle" is a good or a bad thing, but we have already considered how the absence of male-male bonding is a crucial element in the relatively peaceful behaviour of bonobo males. Mothers though, are not exactly the virtuous peacemakers in bonoboland we might expect. On the one hand, male bonding with mothers rather than with other males prevents the kind of violence that depends on male alliances; on the other hand, it brings mothers into the fray.

Kano describes one such affray that occurred in what was then the northern subgroup at Wamba (later named E2), towards the end of his research in 1982. This involved Aki, a powerful female, and her son Koguma who was just entering adulthood, and features adult males Ude and Kuma whom we met earlier. Descriptions like the one that follows help us flesh out the otherwise skeletal understanding we have of the world of the bonobo.

One day, Koguma suddenly, while branch-dragging and screaming, charged violently at the second-ranking male Ude, who leapt up and slapped Koguma. The top-ranking male Kuma calmed Ude by mounting him and then rump-rubbing with him. Koguma charged again, Ude pursued, and they exchanged violent blows. Koguma's mother Aki, with her screaming infant clasping her belly, chased Ude into the bushes.

Koguma carried on charging Ude for a total of 12 times in 9 minutes. When Ude rose to retaliate, Aki vocalised and chased him. After the seventh charge, Ude became silent and avoided Koguma. Subsequently, every time Koguma charged at Ude, Ude fled or he solicited mounting from Koguma which, as Kano says, was an unmistakeably subordinate behaviour.

Kano observed similar behaviours from other males with influential mothers, and he notes how young males seem to be able to judge the potential influence their mothers have, because sons of low-profile mothers do not make these bold challenges against adult males.

Kano also discusses relations between the sexes, saying that, in general, there are two kinds of relationship between males and females: one is sexual while the other is competitive and centres on food. It may surprise the wannabe bonobo to hear that in competition over food, intense hostility and aggression can develop between males and females.

Kano describes how the middle- to low-ranking males were at the head of the party when it first entered the feeding site. A little later the females and the high-ranking males arrived, and the lower ranking males picked up at least one piece of sugarcane and scattered into the marginal areas to avoid aggression from the incoming males. Females and the high ranking males fed together but if there was a shortage of food the males would defend theirs from the females. This defence could involve aggressive behaviour from males towards females, and some females had torn ears as a result of it.

Up to 4 years of age, a young bonobo's behaviour around food was tolerated by adults but then this tolerance disappeared. Juveniles would be chased by older bonobos who would take their sugarcane and threaten them, sometimes holding down the youngster while giving them a bite.

As males and females approached adolescence they spent more time alone at the periphery of the party. The adolescent females mostly kept to themselves but an adolescent male still travelled with his mother, and kept company with her during rest periods. Other than this he too kept to himself because if he was at the centre of the group he would often be threatened or attacked by other males. Kano found that adolescent males rarely mounted and rump-rubbed with adult males, so these males seem to have forgotten the "make love" side of bonobo life.

In a *Natural History* article by Kano we have another description of how things played out at the feeding site. Kano describes the arrival of three males at the site who then moved to the trees, mounting and rump-rubbing. Then the females arrived screaming, and we get ten minutes of chaos before they settle down to feed. During those ten minutes of chaos there was gg-rubbing between females, copulations between males and females, and frequent forced mounting between males, the subordinate male screaming. The sexual behaviours between males were often preceded by aggression or outright attacks, which were always from the dominant male (Kano 1990).

Furuichi's mid 1980s study of the E1 community, covering 97 days and mostly (77.7% of the time) in the natural forest, gives us the following figures for aggression: 87

occurrences of aggressive and/or submissive behaviours between males, 27 between females, and 52 between males and females. In the latter cases the male dominated the female 27 times, and the female dominated the male 25 times; a fairly even split rather than the universal female dominance we might have expected (Furuichi 1997).

Most of these aggressive behaviours were display behaviours (the bonobo stands bipedally with raised arms or a swaying body) or non-physical charges, and only some involved physical contact such as when an individual was grasped, beaten, kicked or bitten by another. Submissive behaviours ranged from simply moving out of the way, to a fleeing escape, to screaming by the victim of the attack.

Aggressive interactions between males occurred most frequently when they became excited on finding a preferred food source or when they heard vocalisations from approaching groups. At these times the males ran around excitedly and showed aggressive behaviour towards other males, though not usually serious physical attacks,

The next most frequent aggressive interactions between males were play-like provocations of dominant males by subordinates, and were very moderate and ritualistic. These often involved late-adolescent males seeking to improve their status, the younger males testing adult males by provoking them. This type of behaviour also occurs in chimpanzees.

The third context of male-male aggression was intentional aggression shown by a male as a way to display his dominance over another. These acts of aggression were most frequently shown by late-adolescent males challenging for alpha status.

For female bonobos, status is related more to age, and the female dominance hierarchy is therefore more rigid and is rarely expressed in aggressive interactions. Most dominance/submissive interactions between females occurred during feeding, and nearly half simply involved the displacement of another female without the use of aggression. There were, though, 7 incidents of physical attack in a battle between females over alpha status. Besides these battling females there was also another adult female, Halu, who frequently and abruptly attacked younger females feeding around her. Halu raised her rank in this period so this behaviour appears to have been an intentional display of power by her.

Aggressive behaviours directed by males towards females were often in the context of display by excited males, with 3 of the 27 cases involving physical attack of the female. Aggressive behaviour by females towards males occurred when females were supporting their sons against other males, or when two or more females allied against a male. Most female-dominant interactions with males involved a simple retreat by the male.

At this time the E1 community comprised 7 adult males, 3 adolescent males, 7 adult females, 3 adolescent females, and 8 juveniles and infants making a total of 28 members. Mothers of 3 adult males and 2 adolescent males were present. As we know, the status of male bonobos can be greatly influenced by their mothers, and the three adult males with mothers in the group occupied high status, though one adult male without a mother was also high-ranking. The other three prime adult males without mothers were lower in rank than the younger males.

The four high-ranking adult males were usually at the centre of the party, and the three middle-ranking males without mothers were usually peripheral. The two low-ranking male adolescents with mothers were also usually at the centre of the party, though they were frequently on the receiving end of aggression from other males.

Furuichi provides some fascinating information on bonobo politics spanning the 12 year period from 1982 to 1994. In what follows, we'll see how the dominance relations between females can also influence those between sons.

In 1982 the oldest female, Kame, had been alpha female for some time, and her oldest son, Ibo, was alpha male. (We met Ibo earlier when, as a young adult, he was assaulted by a stick-wielding older male.) In the autumn of 1983, Kame was pregnant and, probably due to her pregnancy and her old age, seemed less vigorous than usual.

A young male named Ten began to provoke Kame's three sons, Ibo, Mon, and Tawashi, by displaying and making charges towards them. Alpha male Ibo did not show any submissive behaviour, often mounting Ten after these provocations, though he would also sometimes leave the feeding site to avoid them. Kame rarely intervened but, a month later, Ten's mother Sen began to attack Kame's sons. One incident was observed where a number of individuals became involved in a whopping fight, and there was severe battling between Sen and Ibo, with Ibo finally fleeing from Sen.

A few days later a hand-to-hand fight was observed between the two mothers, Kame and Sen. The two females rolled around on the ground, and the skirmish ended with Sen holding Kame down. The defeated Kame continued to scream even after Sen had left the scene. Further fights followed between the two females, and Kame was never seen to defeat Sen who had now become the dominant female.

Initially, Kame's son Ibo seemed to remain dominant to Ten but then, after an aggressive approach by Ten, Ibo, rather than face him off, presented his rear for a rump-rump contact. In January 1984, Sen and her son Ten displayed dominant, branch-dragging displays in the centre of the group while Kame and her sons showed no resistance.

By the time of the next study period in August 1985, Sen and Ten were clearly alpha female and alpha male. Sen's 5-year-old second son, Senta, also began to show very oppressive behaviours: he frequently tried to take sugarcane from adults and, if they refused, he beat their face and head while screaming. If the possessor still refused to release the food, Sen herself came to attack the individual so Senta could have his way.

By the third study period in September 1987, a middle-aged female Halu, and her 10-year-old son Haluo, started to show dominant behaviours. By this time Halu was already dominant to Kame, attacking her several times while Kame showed no resistance. Kame's eldest son, Ibo, was no longer seen and was presumed dead, but her other two sons, Mon and Tawashi, were on the receiving end of repeated aggression from the 10-year-old Haluo.

Halu is the female mentioned earlier who frequently and abruptly attacked younger females feeding around her. In this way, though she was not directly challenging Sen, Halu was able to appear the more dominant female. In November 1987, Halu attacked Sen's son Senta while he was feeding next to his mother. Senta screamed and Halu immediately went behind Sen and lightly mounted her, as if to appease her. The females did not fight but this behaviour by Halu was showing signs of how she was gaining status on Sen.

In February 1988, alpha male (and Sen's son) Ten attacked Haluo (Halu's son) at a feeding site and Haluo fled, screaming. Sen, Halu and other females ran to them and there was another massive fight involving many individuals, at the end of which Halu beat Sen on the head then left the scene. Sen, while screaming loudly, approached Halu and lay in front of her to solicit a gg-rubbing. Halu ignored her, then Kame gg-rubbed with Sen and she stopped screaming.

By the fourth study period in July 1989, Halu's son Haluo was no longer to be seen and was presumed dead. Halu was no longer showing such aggressive behaviour, and Sen and her son Ten were still alpha female and alpha male.

Between the fourth and fifth study periods the old female Kame is believed to have died (she was over 45 years of age when she was no longer seen). By the fifth study period in December 1990, Kame's youngest son Tawashi, who was now about 16 years of age, was on the periphery of the party. Tawashi did not usually enter the artificial feeding site when others were there, and he was chased and attacked if he did. Kame's other son, Mon, who was about 23 years of age, was absent from the group altogether and ranged alone. It is believed that the low status of these males was related to the death of their mother.

By the sixth study period in July 1994, all the old adult females were gone and presumed dead. Halu was the oldest female and probably the alpha female, but she had no grown son. Ten was still the alpha male while Mon and Tawashi remained peripheral community members.

These records of bonobo behaviour over a number of years give us a fascinating insight into bonobo social and political conflict. The study periods varied from two to six months in duration, so we do not have a continuous record of events, but what we do have remains invaluable. Clearly status matters to bonobos, and it matters to bonobos of both sexes. Though males are generally more aggressive, the females can show some of the most severe, if infrequent violence in pursuit of alpha status.

A final thing to note from these observations is the disappearance of the two males from the studied community. Kame's eldest son, Ibo, went missing and was presumed dead when he was about 25 years old. Halu's son, Haluo, was only 12 years old when he went missing and was presumed dead. We don't know how these males died but we have seen how low-status or deposed males can end up on the periphery of a community which could increase their chances of meeting their demise. Whether their deaths were the result of injuries sustained from other bonobos we cannot know but it is a possibility that cannot be completely ruled out.

Kano (1992) observed many handicapped individuals at Wamba, including 46 bonobos with at least one abnormality of the limbs or digits. There were also many examples of abnormalities of ears, eyeballs, genitalia and other body parts. In the cases of finger and toe injuries, there were 28 instances of total loss of the digit, and 96 of partial loss, plus many other abnormalities. Nearly twice as many males as females had physical defects, and of the 166 abnormal fingers or toes, over three times as many occurred in males than in females. All but one of the 22 full adult males seen at Wamba had a defect of some digit, compared to 11 of the 16 adult females.

It is possible that these injuries were due to falls or were sustained in traps or snares but fighting cannot be ruled out either. Ear lacerations, which cannot be blamed on accidents or snares, were seen in 24 males and in 8 females, and these almost certainly resulted from fights.

There were also three cases of the loss of testicles, and one of these males also had no penis. Are these birth defects or had these most sensitive parts been deliberately bitten off? We'll see in a moment why the latter cause cannot be ruled out but, just to complete the picture of bonobo health and fitness in the wild, we'll add that there were also bonobos with possible cataracts, respiratory illness, skin disease (including one case that was particularly disfiguring), and several old individuals had fist-sized tumours on their skin.

Frans de Waal (1997) writes about the injuries that have been reported for captive bonobos due to fighting. Amy Parish recorded zoo cases where females, often collectively, attacked and wounded males. At Stuttgart Zoo a female was always attacking the adult male; she once bit his penis almost in two, and it was left hanging by a tiny piece of skin. The penis was sewed back on by the zoo vet, and apparently recovered its function. At Frankfurt Zoo, the females would sometimes bite the male while holding him down. Parts of his fingers and toes were bitten off, and his testicles were also badly scarred from where he had been bitten many times.

Jahme (2001) writes that the San Diego bonobo males were, on occasion, so terrified of the females that they refused to come in at night for food, preferring to go hungry.

At Apenheul Zoo in the Netherlands, five female bonobos attacked a male and were seen gnawing on his toes; his flesh could be seen between their teeth as they chewed away. And at least two keepers, one at Columbus Zoo and the other at San Diego Zoo, have lost bits of fingers. Jeroen Stevens, a Belgian biologist who has spent thousands of hours studying captive bonobos in European zoos, considers bonobos to have a far tenser disposition than chimpanzees, and to be prone to really violent aggression in larger captive groups (Parker 2007).

It is not only the captive males who have suffered at the hands of the females. In 1999, a female was introduced into the bonobo group at Plackendael Zoo in Belgium. In the first year this female was often the victim of aggression inflicted by the two resident females, who also formed coalitions against her. Even the group's 6-year-old female harassed her frequently. The newcomer showed many signs of distress, such as frequent grinning and pacing, and she was usually found on the periphery of the group.

For the first two years this female failed to show any sign of sexual swelling or menstruation, and though she mated with all the four males she did not become pregnant. In the second year the aggression decreased and she started to show more friendly bonds with one of the females. By the end of that year the female was fully integrated, spending time with other females and feeding near to them. Only then did she resume menstruating and a regular swelling cycle, and after two normal swellings she became pregnant (Vervaecke *et al.* 2003).

This is the same zoo study mentioned earlier where the three adult females rarely mated with the four sexually mature males, only doing so 26, 13, and 8 times in the five-and-a-half month study period. In addition to these copulations there were also 102 non-copulatory mounts and 105 presentations, making a total of 254 sexual interactions. In 23 of these sexual interactions there was aggressive intervention by another adult: 14 by a female and 9 by a male. The dominant female was responsible for 12 interventions, and she appeared to be targeting the female of the pair and not the male. In some instances this was clearly the case, such as her particularly aggressive reaction when the second ranking female was copulating with the lowest ranking male, pulling the female off the male while biting her and chasing her around.

To add to all this less-than-bonobo-like behaviour, there are also cases of aggression by females towards infants of other females. In 2001, the alpha female at Twycross Zoo in the UK took the newborn infant from the lowest ranking female, even though the alpha female was still nursing her own four-year-old offspring. The dominant female nursed the infant for six weeks, which sounds very motherly of her except that, while always a caring mother towards her own infants, she was remarkably rough with the kidnapped infant, often pushing it around or leaving it crying on the floor beside her. When there was group excitement she would even drop the tiny bonobo on the floor. After six weeks she lost interest in the infant which was by then showing signs of weakness and dehydration, and was removed for human rearing.

Acts of aggression by females towards another's infant have also been seen in other zoo populations, including Plankendael and Stuttgart. In the latter incident the dominant female threw a nursing infant against the wall. In all but one of nine recorded cases where an infant was taken or was the victim of aggression, the mothers of those infants behaved nervously and showed signs of distress such as vocalising, grinning, and trying to retrieve the infant. Often, these distressed mothers would present for gg-rubbing to the female who had taken their infant (Vervaecke *et al.* 2003). How sexy is that?

We cannot leave this discussion of aggression amongst captive bonobos without mentioning the rescued bonobos at the Lola ya Bonobo sanctuary in the Democratic Republic of the Congo. The Lola ya Bonobo sanctuary has become well known to bonobo followers, and numerous research experiments are now carried out there, including the one above which looked at the sociosexual behaviour of young orphans (Woods and Hare 2010). Much-needed work is done by the sanctuary to save orphaned bonobos who have ended up as pets after their mothers, and others in their group, have been killed for bushmeat. Many of us will have seen or heard about these adorable bonobos but, as those who tried to keep them as pets discovered, bonobo babies soon become very strong and very unpredictable creatures to have around.

In 2011, one visitor to the sanctuary described how, as adorable as some are, the whole "bonobos are friendly, loving creatures" is a bit of a misnomer. By human standards he found them to be quite aggressive, and they would dish out a very hefty slap or a bite to other bonobos, and to humans who got too close. These were not the peaceful, loving apes this visitor had anticipated meeting but were aggressive, and even their play behaviour seemed malicious.[2]

The visitor describes one incident where an adult male walked by a smaller one in the enclosure and "thud", kicked him right in the stomach. Then he walked by another and beat him, bloodying his mouth while the victim cried and yelled. Even though these are rescued bonobos, and we cannot assume that their behaviour at the sanctuary is quite what it might be under normal conditions, it is important to balance the 'cute apes having lots of friendly sex' image with this other, very real and far less friendly side to their behaviour.

Just as this visitor had arrived at the sanctuary, news was coming in that three trackers had been bitten by bonobos at the sister-site, Ekola ya Bonobo, where they were released back into the wild. The trackers, who were known to their bonobo attackers, were immediately flown to hospital and one of them was released after a month. The other two men were more severely injured having lost their noses, several bits of their fingers, and one had also lost an ear.[3] These two trackers spent much of the following year receiving reconstructive facial surgery in Paris, France. It is clearly not only chimpanzees that are capable of ripping people's faces off.

─────────────────

[2] http://www.spencersekyer.com/lola-ya-bonobo-kinshasa-congo/

[3] http://lolayabonobo.wildlifedirect.org/category/ekolo-ya-bonobo/

From bonobo aggression at Wamba and amongst these captive groups, we'll now move on to take a look at findings from Lomako. Hohmann and Fruth (2003) examined aggression by bonobos and how it related to mating, making the very important point that there is a lot of speculation about bonobo behaviour due to the scarcity of actual data on them. Their study of the Eyengo community comes from data on six field seasons, ranging from two to seven months, and spanning the period from 1993 to 1998.

Community membership at Eyengo ranged from 30 to 36 individuals, with the number of adult females ranging from 12 to 15, while the number of adult males ranged from 3 to 7. As mentioned earlier, this female-biased sex ratio is what we normally associate with chimpanzees, and particularly with male chimpanzee violence and a high male mortality rate due to that violence. The fact that there are also these female-biased sex ratios in bonobo communities means that using such evidence for male violence in chimpanzees is not actually valid.

At Lomako, Hohmann and Fruth found that there was more aggression between males on days when mating occurred than on days when it didn't (yes, bonobos do have days without any sex). The data also showed that when males behaved aggressively they mated more often than when they were the targets of aggression, and mating by low-ranking males was more likely to be disturbed than was mating by high-ranking males. This shows that males use aggression to compete over females, and that aggression pays off when it comes to mating success, even in bonobos.

Unlike the E1 community at Wamba, the Eyengo community at Lomako foraged in smaller parties. Hohmann and Fruth give us some details of the composition of these foraging parties which help us get a better picture of this community, keeping in mind that there were many more females than males.

Mixed parties accounted for 69% of all the parties observed, all-male parties accounted for 8%, and all-female parties for 21%. We also have a breakdown of the composition of the mixed parties: 318 comprised more than one male and more than one female, 215 comprised one male with more than one female, 67 comprised one male and one female, and 24 comprised one female and more than one male.

This is very different from the picture we have from Wamba of a couple of dozen or so bonobos foraging fairly closely together. In particular, the images we have of bonobos at the artificial feeding sites at Wamba have given us a somewhat distorted view of bonobo social life, and findings from Lomako help us adjust our initial thinking to a more natural setting.

At Lomako, as well as aggression between males in relation to mating, the females were also harassing and disturbing other females. This behaviour increased as the number of oestrous (maximally swollen) females present in the party increased, and aggression between females was about 7 times higher when two or more oestrous females were in the party than when there was only one.

Female bonobos (like the males) had higher mating rates when they were aggressors compared to when they were the targets of aggression, and female aggressors were more likely to be in oestrus. The authors of the study also note that females appeared to be competing for access to certain males, so this mating competition between females was about specific males and not due simply to a shortage of them.

Aggression by males towards females was uncommon, and not a successful male mating strategy. What the authors found was that certain unrelated males and females were associating together in the same party, and this association could last for quite some time – from a few months to almost two years. In these close associations between a male and a female there was both a low rate of aggression between them and a high rate of mating. It looks like this male-female bonding was good for a male's mating opportunities and, therefore, his reproductive success.

Male reproductive success was also biased towards high-ranking males so it seems that females were choosing to associate more, and mate more, with such males. We have seen how dominant males can force low-ranking males away from the females, so how much real agreement there is between this male dominance behaviour and the desire of females is not clear-cut. As the females do not appear to do anything to counter the way low-ranking males are excluded by the dominants we must assume the females are happy with the result.

We'll now look at more details of the general aggression that occurred, keeping in mind how few males there were (3 to 7) compared to females (12 to 15).

In mixed parties, 466 aggressive interactions were observed. Of these, 38% involved two males, 23% involved two females, 26% were of female aggression towards a male, and 11% involved male aggression directed towards a female. Of male aggression towards females, oestrous females and younger females were more likely to be targets, and 4 of these incidents of male aggression towards females were in the context of actual mating. This male aggression in the context of mating is therefore rare, but it is not something that never happens amongst bonobos.

The main findings from this study point particularly to the impact good relations with a particular female can have on a male's mating and reproductive success. Many of the bonobo female's sexual swelling cycles are not fertile cycles, and ovulation is a relatively rare event. By associating with a particular female for months, and even up to two years, a male can increase the number of times he has sex with that female, and so he can increase his chances of fertilization with her. As these associations do not appear to be enforced by the male, good relations with the female are important to him.

Looking at how the overall rates of bonobo aggression compare with those of the chimpanzee, it was found that male-male aggression rates are similar in chimpanzees and bonobos. Between the sexes there is more male aggression against females in chimpanzees, and more female aggression against males in bonobos. And though the authors give no details, they do state that fatal injuries are rare but not absent in bonobos. (We'll see one reason for this assertion when we come to an incident witnessed by Gottfried Hohmann at LuiKotale.)

In this Lomako study of bonobo aggression, Hohmann and Fruth also conclude that neither this study nor other data from wild bonobos lend much support to the view that females form alliances to collectively defend themselves against males (Parish 1996). Groups of bonobos were seen to charge resident or stranger males but, while most attacks were led by females, the attacking group included both sexes. Often the males took an active role in these attacks, and their fighting was fierce and violent. Rather than coalitions of females deterring male aggression, it appears that females are supported by both males and females.

When the captive studies had found females joining forces against males there was often only one adult male, and this, of course, would make it appear that females were forming coalitions against males. In the natural conditions of the Lomako bonobos, these coalitions have been found to include both sexes.

And finally, it should be noted that resident males were sometimes seen to charge females with small infants, and this provoked remarkably intense counter-aggression by both females and males against the aggressor. Why a male does this is not known. Female chimpanzees with young infants tend to stay away from the males so it is hard not to see some similarity here between the males of both species (as a potential danger to infants), though for bonobo females such charges do not result in any injury to mother or infant.

There was one observed case of infant death in the Eyengo community but this was when a newborn was kidnapped by a resident female. The mother had tried to retrieve the baby but had failed, and a day later the mother was seen carrying the body of her dead infant. What had happened in the intervening period had not been observed (Hohmann and Fruth 2002, 2003).

Our last port-of-call in our journey through bonobo aggression is the most recent research site at LuiKotale. Following on from the LuiKotale study (discussed above) regarding the influence of mothers on their son's mating success, further studies have been done at this site concerning aggression, mating, and relations between the sexes. This community comprised 9 sexually mature males and up to 16 sexually mature females.

As we saw earlier, when it came to the influence of mothers (6 of the 9 males had mothers present) it was found that this did not affect the mating success of high-ranking males but it did affect that of middle- and low-ranking males. This was mainly due to the proximity to oestrous females sons could gain by staying near their mothers. When it came to aggression, there were 134 aggressive interactions when conflicts arose over oestrous females: 95 involved two males, 37 involved a male and a female, and 2 involved two females. In 30 of the male-female cases it was the mother of the male who was trying to mate who was involved (Surbeck, Mundry, and Hohmann 2010).

Using data from the same study period, the behaviour of males as aggressors was examined (Surbeck *et al.* 2012). All the mature males over 10 years of age were seen to engage in aggressive interactions with other males over access to the females. Males were the aggressors on 577 occasions which was 75% of all aggressive encounters, and males were targets of that aggression 78% of the time, females 22%. Of this aggression by males, 17% involved physical contact and, though females were much less often the targets, the males were more likely to use physical contact aggression against females than against males (32% of the aggression against females vs 14% of that against males).

Female bonobos have sexual swellings even when they are not fertile, but in this study the researchers measured when females were actually fertile by counting back from the times they conceived. What they found was that male aggression increased significantly when females were present who were having fertile cycles, though aggression was not aimed at these females. It would seem that bonobo males do have some awareness of when females are most likely to conceive. It is possible, even probable, that female behaviour changes when she is ovulating and the males can pick up on this change in behaviour.

Overall, high-ranking males had higher aggression rates than low-ranking males, and males with higher aggression rates also had higher copulation rates with oestrous females. Males with higher ranks also groomed unrelated females more frequently than did lower-ranking males, and these friendly associations with females reduced male aggression and increased mating between the pair.

So this is the same as was found at Lomako. As at Lomako, high male rank allowed proximity to females and therefore mating opportunities. These high ranking males had good relations with these females and, if we also conclude that the females could have chosen to associate with lower-ranking males if they had wanted to, it appears that the females prefer associating with these dominant males.

In 297 conflicts between females and males observed at LuiKotale, females won 56%, males won 35%, and the remaining 9% were undecided. The female initiated the aggression in 54% of these conflicts and the male initiated the aggression in 46%. Of these conflicts between females and males, 40% occurred in a feeding context, 9% in a mating context, and 37% occurred in a social challenge context. Of those that occurred in the context of feeding, it is noteworthy that females were not more likely to win these conflicts.

Further conflict between the sexes occurred when females were giving support to others, and females never lost conflicts in this context. This type of female aggression was about supporting offspring: support was given to sub-adult males 22 times, juvenile non-dependent offspring 11 times, and dependent infants twice (Surbeck and Hohmann 2013).

Neither the number of female party members nor the presence of a close female associate or coalitionary partner had any influence on the outcome of conflict between the sexes. Coalitions between females during conflicts were seen 26 times but there were also 25 male-female coalitions, and 7 coalitions between males. Of 136 acts of aggression by males towards females, only 2% led to a female coalition against the male. This is not what we expect to find when we have been led to believe that there is a bonobo 'sisterhood' keeping such male aggression in check.

Overall, female alliances were rare, and they were almost never provoked by male aggression towards females. They were more often caused by males charging at immatures, and in all cases females stopped the male aggression. The immature targets of male aggression were mostly independent offspring over 5 years old, so this is not about potentially infanticidal behaviour by males.

Across all studied groups of bonobos, aggression is clearly a significant part of their behaviour. Compared to chimpanzees, male aggression is less severe but there is also much more female aggression. Female bonobos can be the most high-ranking, dominant individuals but all females are not dominant over all males. Younger females who are not yet mothers or are only first-time mothers are low-ranking individuals who can often be targets of aggression from others, both male and female.

Frans de Waal has often promoted the idea that bonobos and chimpanzees are as different as night and day (de Waal 2005), but even he has also been willing to acknowledge the fallacy of such a dichotomy. Bonobos, de Waal writes, use dominance and aggression to settle conflicts, as do chimpanzees. The level of violence amongst bonobos may be lower but they are not entirely peaceful, and "the difference from the chimpanzee is slight" (de Waal 2001).

We have seen throughout this chapter that bonobos of both sexes can behave aggressively – and sometimes quite violently – as they compete for mates, food, and social status. Just as bonobos are not all about peaceful living, neither are chimpanzees all about violence. Both species are social animals with complex and largely successful ways to manage social living. And the differences between them are not as great as we have often been led to believe.

Chimpanzee males are relatively strongly bonded, cooperating to protect their community and sometimes to hunt but, very occasionally, the competition between males within the community can lead to fatal injuries. Though direct evidence of a fatal attack within a community of bonobos has not been witnessed, much of what we have seen points to the potential for fatal injury to occur, and some researchers believe this has, in fact, occurred.

One such researcher is Gottfried Hohmann. An example he writes about is an incident when a male jumped on to a branch and appeared to provoke a young female with a baby. The female lunged at the male who fell to the ground. Other females and males jumped down onto the male, creating a scene of frenzied violence accompanied by constant loud shrieking. After thirty minutes of this shrieking violence the bonobos all went back up into the tree, hardly recognisable with their hair on end and their faces changed. Though no blood could be seen at the scene of the attack, the male victim was gone and was not seen again. Hohmann believes the male suffered fatal injuries in this attack (Parker 2007).

From what has been observed of bonobo aggression, whether by males, females, or the two sexes working together, we should not be surprised if their sociosexual behaviour sometimes fails to prevent serious injury, and even death.

We now leave aggression within bonobo communities and turn to the behaviour of bonobos when they meet up with their neighbours.

Five: intergroup behaviour

When we create a mental picture of neighbouring communities of bonobos meeting in the forest, what do we see? Chimpanzees, we imagine, are angry and violent neighbours, out to kill at the first opportunity. But not the sexy bonobos; for these "make love not war" cousins of ours it is simply a welcome opportunity to have sex with the neighbours, isn't it?

Intercommunity behaviour represents, we believe, a major difference between bonobos and chimpanzees: chimpanzees from different communities are inevitably at war while bonobos will happily intermingle. We do have stark evidence that chimpanzees will carry out fatal attacks, especially if they come across lone, vulnerable individuals from other communities, but even amongst chimpanzees these fatal attacks are rare overall. What do we really know about bonobos?

Kano's earliest observations at Wamba were of a few minor skirmishes between two neighbouring communities when they twice shared a main food source in overlapping territory (Kano 1984). Not exactly friendly behaviour but "minor skirmishes" doesn't sound too bad. However, in addition to these, Kano reported that one of these two communities also had an encounter with a third community which was violent. In this violent encounter the intergroup fighting resulted in serious injuries to several individuals – a definite blot on the bonobo intercommunity landscape.

The home range of a bonobo community can overlap extensively with that of its neighbours but rarely did Kano observe Wamba bonobo communities to have direct contact. He describes, for example, how K community, numbering 70, were approached by B community, numbering 30. The B group climbed trees at about 500 metres away and began a loud chorus, and the two groups avoided contact (Kano 1992).

Sometimes, the larger group in such encounters pursued the smaller group. Kano describes one episode where a party from E community was at the feeding site when another party was heard in chorus nearby. The E party members immediately responded with loud vocalisations and left the feeding site, heading towards the sounds. The clamouring voices continued for two hours but there was no contact.

In general, Kano says, bonobos prevent direct encounters by simply avoiding each other. If two foraging parties from different communities are nearby, the smaller party will move away and the larger party gets whatever food source is there. Traveling in a large foraging party means there is more feeding competition within the party but it has this important advantage when it comes to intercommunity competition for a particular food patch.

Early evidence from the bonobos at Lomako showed similar behaviour there (Badrian 1984). At Lomako, intercommunity encounters were also found to be infrequent because vocal contests between foraging parties from different communities led to avoidance of actual contact.

A description is given of one occasion when more than 15 members of the Eyengo community at Lomako were feeding. A smaller Bakumba group was traveling towards them and, while still some distance away, they began vocalising. The Eyengo group responded with great agitation and immediate vocalisations of their own. Both groups continued to vocalise for about twenty minutes but the Bakumba group came no closer. Eventually, the Bakumba group became silent and were presumed to have left. The Eyengo group continued vocalising, and they still showed signs of agitation some 30 minutes later.

Observations from these early years have not yet given us any examples of the fun and sexy get-togethers so many wannabe bonobos are expecting to find. Though bonobo territories overlap while chimpanzee territories do not, these types of vocal contests between bonobo communities are also the way chimpanzees mostly interact with their neighbours.

But then, something different was observed at Wamba. In 1986, the P community at Wamba started to use the artificial feeding sites. On 19 days during a two month period from December 1986 to February 1987, there were 25 encounters between the E1 and P communities at the feeding sites. The number of E1 members present ranged from 25 to 32 (average 28.8), and the number of P members ranged from 6 to 38 (average 21.8). The one case where there were only 6 members of P community was a group of three mothers with their offspring, and the encounter lasted only five minutes. The other 24 feeding site encounters ranged from 29 minutes to 373 minutes, with a total encounter time of almost 42 hours (Idani 1990).

All these encounters between the E1 and P communities began with an exchange of loud barks and increased tension, sometimes leading to aggressive chases and threats, but in time the bonobos settled to feed and rest together. It should be noted though, that the feeding site measured about 1500 square metres, and individuals were well-spaced, often at 15 metres or more while feeding.

What about the intermingling – especially the sexual intermingling – we hear so much about?

There were a total of 26 cases of gg-rubbing involving 8 E1 females and 5 females from P community. One particular female from P community was involved in 15 of these 26 gg-rubbings, while the highest number for an E1 female was 7. The rest of the females of both communities ranged from 0 to 5 gg-rubbings.

So we have a total of 26 instances of gg-rubbing in 25 encounters – an average of one per encounter. On the one hand, this intermingling of females from different communities is noteworthy; on the other hand, we should not be misled into thinking there was much more of this behaviour going on than actually occurred.

The males, in contrast to the females, kept their distance from each other, and any approach was met with aggressive behaviour such as charges or threats. There were 29 aggressive interactions between males of the two communities: 21 of these were aggressive behaviours by E1 males towards P males, and 8 were by P males towards E1 males. Unlike the aggression between males within communities, there were no appeasement behaviours such as mounting or rump-rump contact between males from the two communities. So there was rarely any physical aggression but neither was there any contact between the males that could be viewed as in the least bit friendly.

As for sex between males and females of the two communities, there were 43 intercommunity copulations. This is an average of one per hour and fewer than two per encounter. Most of these copulations occurred within the first 15 minutes of an encounter, immediately after the initial antagonistic interactions but when all individuals were still excited.

This intercommunity sex is again, like the gg-rubbing, noteworthy behaviour but when we consider the numbers of sexually mature males and females, the number of encounters, and the 42 hours spent together, the 43 copulations is far from the orgy of sex many of us have been led to imagine.

Not all bonobos engaged in intercommunity sex. Out of a total of 23 sexually mature females, 4 were involved in 30 of the 43 copulations. These 4 females were: one P adolescent, a recent immigrant, who copulated 10 times, one P adult female who copulated 11 times, and two adolescent E1 females who copulated 5 times and 4 times.

Two prime adult E1 males at the centre of the feeding site each mated 9 times, and one other E1 male mated 5 times. The number of copulations for all other sexually mature bonobos of both sexes ranged from 0 to 3 over this whole period, with 17 individuals scoring a zero for intercommunity sex. There were also 4 cases where E1 males rushed to stop sexual invitation displays by P males.

One example of sex involving an adolescent female of P community is given. This female entered the feeding site after all the sugarcane had been taken, and she approached two feeding E1 males at the centre of the site. She presented to one of these males who copulated with her for 6 seconds then he sat down to continue feeding on the sugarcane that he had held in his hands the whole time. She then presented to the other male who, also keeping his sugarcane in his hands, copulated with her for 25 seconds. The female then snatched a piece of sugarcane from the male and quickly returned to the forest. This sounds a lot less exciting than the fun-fuelled, sexy get-together we might have pictured.

There were 13 instances of aggression between males and females from different communities, 7 of them directed by a male towards a female, and 6 of female aggression towards a male. The male aggression towards females was mild but that of females towards males was fierce, often involving a lot of chasing followed by biting or a beating. There were also 3 cases of intercommunity aggression between females.

As well as these encounters at the feeding site there were also 7 encounters between Wamba bonobo communities in their natural habitat. In three of these the groups barked at each other and there was no direct interaction. In one case between the E1 and B communities there was barking and aggression including chasing and beating and "pressing down" of individuals. Harsh cries and screams were heard and, though not directly observed, a fight was believed to have occurred on the ground. In the remaining three natural habitat encounters the interactions were friendlier. Two of these were between the E1 and E2 communities (E1 and E2 were previously a single community) and the third was between the E1 and B communities.

Idani gives a brief description of this more friendly encounter between 28 members of E1 and 10 members of B community. At the start of the encounter, the bonobos from the two communities confronted and barked at each other. After a while they calmed down, and three gg-rubbings and two copulations were observed between the two groups. The bonobos were together for nearly two hours before the B community members began to leave and move south.

An adolescent female from B community was seen to still be with the E1 group, and she performed a series of three gg-rubbings with an E1 adolescent female. So far, so good. But then, shortly after these gg-rubbings, most of the E1 females joined in an intensive attack on the B female who ran around screaming then left in the direction of where B community had gone. We don't know why the E1 females turned on the adolescent but she was clearly no longer welcome.

The behaviours that occurred during all of these encounters look like an extension of normal bonobo behaviour around food, except for the completely unfriendly relations between males of the two communities. Adolescent females use sex in their visits to other communities, and they also use sex when seeking to establish a permanent residence in one of them, so their relatively high frequency of sex with 'strangers' is not something completely at odds with what we might expect.

Established adult female residents will have had some history of contact with neighbouring communities either as their natal community or as a visited community when they were adolescents on the move. The particularly friendly behaviour of one P community female does not seem so extraordinary when we learn that she apparently grew up in E community (Wrangham and Peterson 1997).

The existence of the feeding site was a major factor in these encounters. The easy availability of abundant desirable food attracted many bonobos to the site, and their presence in such a large open space, rather than the dark forest, perhaps also enabled females to recognise known individuals in the other community.

These were large parties of bonobos from the different communities, averaging over 20 members, and the feeding area was large and well-stocked. While it is significant that the bonobos did share the same food source, as large as it was, we should also note that even chimpanzees from neighbouring communities will not come to blows when parties are large on both sides.

What about the bonobos at Lomako? In an interview with Frans de Waal in *Bonobo: The Forgotten Ape* (1997), researchers Hohmann and Fruth say that the initial intercommunity encounters they observed at Lomako were violent and aggressive, the males wildly chasing each other in the undergrowth, the females hanging in trees screaming. In later observations the bonobos were sometimes seen to settle down and there was some sex between communities. Grooming sometimes also occurred but it remained tense and nervous, and no friendly contact was seen between males. Often, though, communities were still not observed to get along, and there was nothing beyond aggressive displays and loud vocalisations.

Later, in October 1997, a different kind of encounter was observed at Lomako. In this encounter, two strange males appeared in the home range of the Eyengo community. Whenever these strangers ate or rested nearby they were charged by resident males and females. The two strangers acted submissively most of the time but the attacks by the resident males involved severe physical aggression (Hohmann *et al.* 1999).

This severe aggression is a big step up from the more usual aggressive displays of males when two parties from different communities had met. With only two males from another community, and with the odds in their favour, the resident males were more willing to unleash physical violence against them. When chimpanzees kill stranger males it is also when the odds are in their favour, and often when that stranger is alone. Though these bonobo males did not kill the strangers it does show that, should they come across an isolated individual or two, such an outcome is not out of the question.

Of 23 intercommunity encounters at Lomako, 20 included aggressive displays, and 8 of these involved physical aggression. The three that didn't involve any aggression at all were instances of all-female parties (of 2, 3 and 5 adult females) encountering a mixed-sex party from another community. These females showed signs of fear (screaming, hiding) and escaped by running away on the ground, remaining silent and highly vigilant for the rest of the day. In these three cases, one or more females carried small infants, and it would appear that the females considered themselves to be in potential danger from the strangers.

It is also notable that the size of foraging parties increased on the day following an intercommunity encounter, and this was due to an increase in male members. Most encounters between bonobo parties from different communities occurred in large fruit patches, so having more party members enabled better defence of that food. The fact that it was an increase in male membership of foraging parties also suggests that more effective defence and better protection was being provided (Hohmann and Fruth 2002).

Almost all we know about bonobos comes from Wamba and Lomako but we do have a little bit of information from another population of bonobos at the Lukuru research site. Intercommunity encounters at Lukuru were only observed around a series of perennial pools where bonobos regularly gathered to feed on sub-aquatic vegetation, sometimes in large parties of 18 to 25 members. These parties from different communities did not mingle but simply avoided each other. Larger groups vocally announced their approach while more than 15 minutes away which provided the smaller group with the opportunity to leave before they arrived. On two occasions the smaller group did remain along the shore of the pool but they hid in the vegetation, quietly observing as the larger group entered and fed (Myers Thompson 2002).

What about chimpanzees?

Chimpanzees are known to sometimes kill other chimpanzees, and the victims are usually isolated individuals from other communities. These are not everyday occurrences though, and it was many years and thousands of hours of observation before this murderous behaviour was seen. In chimpanzees, like bonobos, vocalisations usually lead to avoidance. If chimpanzees do come into contact the worst it normally gets is aggressive charges between males, and only occasionally is there physical contact. Killings are rare because lone individuals usually avoid the border areas between communities (Wrangham and Glowacki 2012).

The main difference between bonobos and chimpanzees is that bonobo ranges do overlap whereas chimpanzees are much more territorial. Male chimpanzees are strongly bonded and will put aside internal quarrels to form all-male groups that patrol the borders of their mutual territory. Bonobos do not form these male patrol groups, and they forage more often in mixed-sex parties. While communities of bonobos are just as distinct as are communities of chimpanzees, the areas where they forage are less distinct. Bonobos do, though, attempt to defend food patches and, as we have seen, larger parties displace smaller ones.

Just as the severe aggression shown by the male bonobos towards the two stranger males warns us that fatal injuries are not out of the question should they come upon an isolated individual or two, some chimpanzee behaviours suggest that our polarisation of the two species should be reduced from their side too. For this we need to return to the chimpanzees of the Taï forest.

We have already noted how Taï females, especially young ones, have sexual swellings for a much greater part of the interbirth interval, and therefore show great similarity with bonobos. Taï chimpanzee females are more gregarious, and males and females travel together more. Taï females also can have close friendships with each other; one female, for example, immediately adopted the son of her friend when she died (Boesch 2009).

Christophe Boesch observed the North community at Taï for 29 years without seeing a fatal attack, even though there was regular patrolling, and intercommunity encounters averaged about one a month. There were two fatal attacks by the South community at Taï, and the South community also killed an infant. This infant had been left behind when its group members had fled so it had not been specifically targeted. It seems that infanticide is not something the Taï males are interested in because there have been 41 observed cases where Taï mothers have been temporarily captured by members of other communities but their infants have been left unharmed.

Boesch has also observed a few cases where back-and-forth attacks between similarly sized groups from different communities have quietened down and changed to displays. What is even more unexpected is that some females mixed with males from the other community. He describes one occasion where the males formed two parallel lines about 15 metres apart. One group was joined by two young females in oestrus and a mother with an infant. While the two lines of males remained facing each other, these three females walked quietly to the other side and mated with some of the stranger males. At the same time, a female from that group also crossed and mated with her 'enemy'. After some minutes the two groups parted.

This is far from what we would expect of chimpanzees, and it again shows that the clear distinction between bonobos and chimpanzees is not so clear after all, even in this intercommunity behaviour. It is probably significant that these were meetings of two similarly sized groups of chimpanzees where neither could expect to win against the other. But it is also far from expected behaviour of chimpanzees for females to brazenly mate with 'enemy' males in the presence of males from their own community. It should make us stop and think before we draw lines between the behaviour of the two species; the gaps in our knowledge of both these species are greater than we realise.

Our picture of chimpanzees has mostly been formed from research at Gombe and other similar eastern populations. But most chimpanzees live in central and western forests, and the more we learn about other chimpanzee populations, the more we discover how varied their behaviour can be. We still do not know all there is to know about chimpanzees, and we know far less about bonobos, so we should not be surprised when we find such behavioural overlap between them.

For our human wannabes, the release of the inner bonobo has been about sex as a panacea for all human evils. The image of bonobos from different communities engaging in lots of sex rather than conflict seems particularly attractive compared to the horrors of warfare but the reality, as we have seen, is a lot less sexy and a lot more strained.

For both chimpanzees and bonobos it does not look like the males from different communities are ever going to get along. Females of both species transfer to a new community, and visit many, so it is not surprising that their reactions to encounters with other communities are different from those of the males.

For humans, the movement of a young daughter to a new home of strangers, perhaps rarely seeing her birth family again, would be (and is, when it occurs) an enormous disadvantage in life. Bonobo and chimpanzee females choose to make this difficult transition but it is still a stressful experience for the young females of both species, though the adolescent bonobos are not going to be faced with quite the same level of male dominance as are the females of their sister species.

The status and influence of mothers in bonobo society and the absence of alliances between the males, are two major factors in the relatively less violent and domineering behaviour of bonobo males. The mixed-sex parties of bonobos, foraging in areas used by more than one community, creates something very different from chimpanzee territories patrolled by parties of strongly bonded males.

The presence and influence of bonobo mothers may be, as Kano says, an obstacle to male bonding but it has been the gateway to a reduction in the particularly unpleasant consequences of male alliances both within and between communities. As de Waal puts it, "for bonobos, an every man for himself system paves the way for a collective female power takeover" (de Waal 1997).

If we are seriously looking for the reasons why bonobos are less violent than chimpanzees then we need to reduce the focus on sex, and turn our attention towards the most powerful factors: the weakness of male-male bonds, the strength of male-female bonds, and the power and influence of mothers.

Six: the naked bonobo

*F*ew of us will have escaped hearing how bonobos "make love, not war", or how they engage in constant sex in all age/sex combinations, or how they are from Venus and chimpanzees are from Mars. But these soundbites have taken on a life of their own, becoming further and further removed from reality. We still know relatively little about bonobos in their natural habitat, and most of that information is based on only a few sets of data from a few study sites. What this information tells us though, is that their lives are much richer and much less black-and-white than those soundbites have led us to believe.

One unfortunate consequence of the recent bonobo rise to stardom has been that some people even think we are more closely related to bonobos than to chimpanzees. We're not. The three of us share a common ancestor that lived about

5 or 6 million years ago. After our evolutionary paths went their separate ways, the chimpanzee/bonobo ancestor did not split into those two species (separated by the Congo River) until about a million years ago. It would also be a mistake to get too carried away in our thinking about what our relationship to these apes means for us today, when there have been 5 million years of evolution on both sides of the divide since our common ancestor.

Last Common
Ancestor

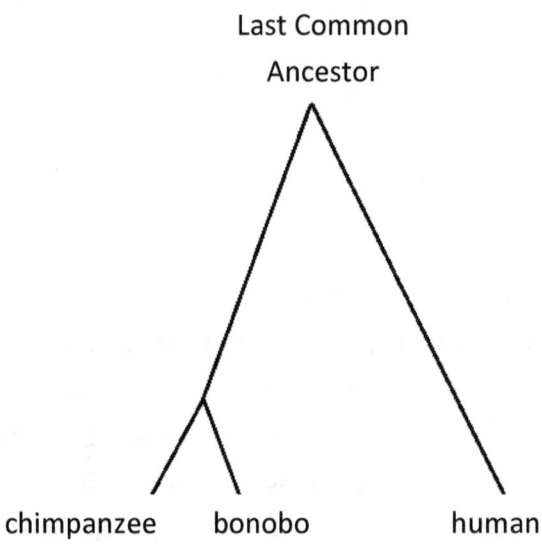

chimpanzee   bonobo   human

It is important that everyone who knows about the chimpanzee should also know about the bonobo, and know that these two species are equally close to us in evolutionary terms. Many other species evolved in our own hominin lineage since we shared that common ancestor but they are all extinct (these relatives of ours include various species of *Australopithecus, Homo habilis,* and so on). Fossil remains are few, and they would tell us little about behaviour anyway, so whether we like it or not, the chimpanzee and the bonobo often get the most attention when we attempt to understand where we have come from, and even who we 'naturally' are.

How does the human genome compare to those of the chimpanzee and the bonobo, or the gorilla and orangutan for that matter?

Bonobos and chimpanzees are very closely related to each other, as we would expect, sharing about 99.6% of their genome. Looking at the number of DNA base changes, humans and *Pan* (chimpanzees and bonobos) are estimated to share about 98.7% of their genomes, and not far behind, humans and gorillas share about 98.3%. As for orangutans, we share about 97% of our DNA with these Asian great apes (Harris 2015).

What is even more interesting is that if we compare the three genomes of human, *Pan*, and gorilla, the human genome comes closer to that of bonobos and chimpanzees only across about 70% of its content. About 15% of the human genome has been found to be more like that of the gorilla than *Pan*, and for the other 15%, the chimpanzee/bonobo genome is more like that of the gorilla than the human genome. In other words, for about 30% of the genome either we or *Pan* are more like the gorilla than each other.

Some human DNA is not even particularly close to either *Pan* or the gorilla: about 1% of our genome is more closely related to the orangutan than any of the African great apes. As we look to release our inner bonobo we should not forget our inner chimpanzee, our inner gorilla, and even our inner orangutan, though we'll have to watch we don't just become terribly confused, with these multiple ape personalities jostling around inside us.

If we just compare humans with chimpanzees and bonobos, more than 3% of the human genome has been found to be more closely related to either the bonobo or the chimpanzee than these are to each other: about 1.6% of the human genome is more closely related to the bonobo than the chimpanzee genome, and about 1.7% is more closely related to the chimpanzee than the bonobo genome (Prufer *et al.* 2012).

What does any of this tell us? On the one hand, we are getting a better picture of the genetic similarities between the apes; on the other hand, the findings are not as straightforward as we might have hoped. For one thing, it appears that our inner gorilla needs a lot more attention than it has been receiving of late. We also have to consider the complexities of genes, how they interact within the genome, and how they interact with the environment. So, taking all this into account, we still don't really know much at all.

It would not be too difficult to argue that everything we discover about other apes is ultimately meaningless in our quest for a better understanding of humans. So much has happened in our own evolution, both physically and culturally, that it places us worlds apart. But we do share relatively recent evolutionary roots, and despite our species' particularly complex features we are still 'just' another evolved species, and not one that suddenly appeared on this planet as a readymade and supernaturally spawned lifeform. Our primate connections can still be very apparent when we spend any amount of time watching our ape or even monkey cousins.

With the bonobo currently at centre stage, we need to remember that she is no more closely related to us than is the chimpanzee – and we should especially remember that the bonobo and chimpanzee are the species most closely related to each other in this trinity. No matter how hard we try to paint these two species as polar opposites, the fact remains that they are very closely related (both are clearly chimpanzees), and humans are far more distantly related to either of them than they are to each other.

But now it is time to look at those bonobo soundbites and to remind ourselves what the reality is.

Bonobos are matriarchal/female dominant

What is meant by the term "matriarchy" when it is applied to bonobos?

When we use the term "patriarchy" for humans we are talking about male dominance in all spheres of life, and about the father as the head of the family, passing on power, influence, and wealth to his sons. If we consider chimpanzees to be patriarchal it is because the males are dominant but there is no particular relationship between fathers and sons, who are ultimately unknown to each other anyway. Males stay for life in their birth community so we do have a collective continuity of home territory between all fathers and all sons in this sense, but the "patriarchy" of chimpanzees really just means that males are dominant over females. And clearly so.

Male bonobos also stay for life in their birth community so in this respect they are no different from chimpanzees. When people say that bonobos are matriarchal they often simply mean that females are the dominant sex. Sometimes there can be more of an emphasis on females as mothers, and this emphasis on mothers and maternal behaviours adds to the image of bonobos (in opposition to chimpanzees) as nurturing and kindly creatures.

For the most part, all adult male chimpanzees are dominant over all female chimpanzees, but bonobos are not a simple reversal of this. The most dominant individual or individuals can be female but beyond that, some males dominate at least some females. Males do not simply show submissive behaviours towards females in the way female chimpanzees are submissive towards males.

Frances White, who carried out some of the main studies of the bonobos at Lomako, has even found that all adult and sub-adult males she observed out-ranked all females. She also found that male aggression towards females was relatively common, and male submission to female aggression was rare. Females did have priority when feeding but this was either because males were too busy trying to exclude and evict each other from the fruit tree, or because males would rest for a while, allowing the females to feed first. In neither case, therefore, was this feeding priority the result of female dominance behaviours leading to the displacement of males (White and Wood 2007).

Research studies vary in what they find because there is not the clear female dominance we might have expected.

Some researchers consider females to be more dominant, some consider males to be more dominant, some think there is co-dominance, and others simply say that dominance relations are ambiguous.

Male and female bonobos each have their own hierarchy, and an individual's position within their sex hierarchy affects how likely they will be able to dominate a member of the other sex. Young females in particular occupy low status and keep a low profile. It isn't until a female has produced at least a couple of offspring that she improves much on her lowly position.

We have seen how mothers with growing sons can be out to gain power and status, and how this improves the status and mating prospects of those sons. Kano observed how young males seemed to be able to judge the potential influence their mothers had, because the sons of high status mothers made bold challenges against adult males whereas sons of low status mothers did not.

Maybe it is the prospects of a son that motivates a female to seek higher status. We have seen how behaviours of mothers such as Aki, Sen, Kame, and Halu at Wamba were interlinked with those of their sons. And we cannot say that it is simply age that leads to increasing female dominance because we have also seen how the weakness of Kame in her old age led to a takeover by a younger female.

Frans de Waal commented that females have nothing to compete over except their sons' careers. This may well be a major part of bonobo life for females, though we would expect high status to also be very useful to the female herself when it comes to access to limited resources, and we have seen examples of this in the way dominant females will displace subordinate females from food.

At Wamba, Kano described how, when it comes to food, there can be intense hostility and aggression between the sexes, and some females ended up with torn ears. While at LuiKotale, 40% of conflicts between the sexes were over food, and females were not more likely than males to be the winner. Males winning conflicts over food is something that, according to captive studies, is not supposed to happen in bonoboland. Yet it does, and even the image of females always having priority of access to food needs revision.

The debate about which sex is dominant or whether the sexes are co-dominant continues because dominance relations between the sexes are not clear-cut. Whereas male dominance is clear amongst chimpanzees, the relations between the sexes in bonobos are more complex and mixed. But at least we can say that bonobo females do better than their chimpanzee sisters.

What needs to be kept in mind is that the main beneficiary of this female dominance, and maybe the motivation for her behaviour, can clearly be a male, even if it is her own son. Daughters leave at puberty and that is the end of the mother-daughter relationship, but mother-son relationships last for the mother's lifetime. In terms of a female's reproductive fitness (which is ultimately how behaviours evolve), mother bonobos succeed in leaving more grand-offspring by aiding and abetting sons, and this creates a more complex interaction of male and female interests.

Female bonobos can be the dominant individuals but all females are not dominant over all males. Young immigrant females who are not yet mothers or are only first-time mothers are low-ranking individuals, and they are often targets of aggression from other bonobos of both sexes.

Compared to chimpanzees, female bonobos do fare much better, especially the established mothers. "Matriarchy", in the sense of the high social position of some mothers rather than a general female dominance, would seem to be a reasonable description of bonobo life, though the jury is still out.

Bonobo females are strongly bonded in a 'sisterhood' that acts cooperatively to dominate males

A powerful image we have of bonobos is one where all females have strong and egalitarian bonds of friendship, and if any male steps out of line they collectively put him right. This is an impression gained mostly from early observations of a few small, captive groups, where there was often only a single young adult male. We have learned that in natural conditions, things are different.

Research at both Lomako and LuiKotale has found little support for the view that females form these alliances to collectively defend themselves against males. At Lomako, bonobos were seen to collectively attack an aggressive male but these coalitions, while led by a female, often contained both sexes.

At LuiKotale, 26 coalitions between females were seen but there were also 25 male-female coalitions, and 7 coalitions between males. Of the 136 acts of aggression by a male towards a female, only 2% led to a female coalition against the male. This is clearly not females acting together to protect each other from male aggression. Female alliances were mostly the result of males charging at immatures, most of these being independent offspring over 5 years old.

Though females may act together against an aggressive male this is only part of the story, and males are just as likely to be acting alongside females. Female bonobos do not form a 'sisterhood' whereby males can be consistently dominated, nor do female alliances act to deter males from aggression towards mature females. Relationships between male and female bonobos are multifaceted, and not as simple as the soundbites lead us to believe.

Bonobos are not violent
Compared to chimpanzees, bonobos are less violent but that does not mean they are non-violent. Much of their aggression does stop short of physical contact but not always, and neither are females immune from attacks by males or from being the violent aggressor.

Most bonobo aggression is between adult males, and at the artificial feeding sites Kano told us that threats, chases, and attacks were so frequent that some bonobos could not get any share of the food. This type of behaviour is not restricted to the artificial feeding sites because males at Lomako have similarly been denied access to food in fruiting trees: two pubertal males, for example, only got to eat when they were traveling alone with their mothers (de Waal 1997). One of these males disappeared during the study period, and though the cause of his disappearance is unknown, such poor treatment of young males in a bonobo community is not going to be helpful to their survival.

At Lomako, Frances White observed two occasions of sexual coercion, while Hohmann and Fruth reported four cases of male aggression against females in the context of mating. It is said that "rape" does not occur in bonoboland but this evidence shows that it occasionally does.

Between males, mountings and rump-rubs are often forced on a shrieking victim; as we are calling these male-male behaviours *sexual* behaviours, then we must call this behaviour sexual assault. Because penetration does not occur in this male sexual coercion we cannot call it "rape", but we should note that bonobo males are, nevertheless, regular victims of sexual assault. Wannabes should be thankful that penetration is not a feature of male-male 'sexual' behaviour; if it were, rape of males by males would be a defining feature of the bonobo, especially at the artificial feeding sites.

We have been treated to the fascinating account from Wamba of the violent battles involving mothers (Kame, Sen, and Halu) and sons (Ibo, Ten, and Haluo). Though males are generally more aggressive, the females can show some of the most severe, if infrequent violence when in pursuit of alpha status or in support of their sons.

Female violence, unlike that of males, often occurs without warning. Kano (1992) writes, for example, how on arrival at the feeding site a female would sometimes jump on a neighbouring female, drag and bite her, hold her down, and steal her sugarcane. Not very sisterly behaviour at all.

Females have also shown behaviours towards infants, such as kidnapping and maltreatment, which could potentially lead to the death of the infant, and in at least one case did lead to infant death.

We also noted that two Wamba males, Ibo and Haluo, were "missing presumed dead", and how we cannot rule out bonobo violence as a potential cause or contributory factor in these deaths. Kano recorded very high numbers of injuries and deformities amongst the bonobos at Wamba, at least some of them due to fighting. Researcher Gottfried Hohmann does believe that there have indeed been fatal injuries amongst bonobos, and he describes one attack on a male community member who was never seen again, as well as severe attacks on two male strangers.

When we add up all the evidence from natural communities, zoos, and Lola ya Bonobo, we do not find a non-violent species; nor would it come as a complete surprise should we ever get clear evidence of a bonobo fatality resulting from bonobo violence.

Overall, the rate of aggression between males is similar in chimpanzees and bonobos. Between the sexes there is more male aggression against females in chimpanzees, and more female aggression against males in bonobos. It has been assumed that the lower frequency of physical violence between bonobo males is due to sex with females being more readily available and equally shared, so that there is little or no male sexual competition, and little reason for any male to feel aggrieved. This no longer seems to be the case. We have found that aggression between males pays off when it comes to their mating success, and that females appear to prefer dominant males as their mates.

There is growing evidence that lower ranking males employ self-restraint or are secretive in their pursuit of females, so male sexual competition is often subtle rather than absent. Bonobo males are not getting equal access to sex, and they are not indifferent to the sexual activity of other males. It appears that the lower level of male violence is a consequence of males who are more willing to accept their place rather than strongly challenge more dominant individuals.

The figures have shown that mating and reproductive success of males is linked to their rank in the male hierarchy. It is still not clear how males establish their rank but they do not form the kinds of male alliances formed by chimpanzee males in pursuit of power and status. The benefits that can be gained from good relations with unrelated females, and from close relations with mothers, appear to have outweighed the potential benefits of alliances between males.

When it comes to intercommunity encounters, it is true that there have been no observed killings. When these occur in chimpanzees it is usually because a lone individual has been discovered by a group of males from a neighbouring community. We do not have observations of this kind of encounter of a lone individual in bonobos but we do have the observations of severe aggression against a pair of stranger males. This illustrates the potential amongst bonobos for violent attacks of vulnerable strangers, and warns us that fatal injuries in such encounters cannot be ruled out.

When we looked at the intermingling of bonobos from different communities, most of these were during a specific and isolated two month period, and were at the artificial feeding sites with large numbers present from each community. Some females did cross over to interact but males remained at an unfriendly distance. The intermingling by the females turns out to be not so strange when we remember it is females who transfer between groups, and interaction with strangers is a normal part of their life. Some of these females will not even be interacting with strangers after all, but with members of the community in which they were born, or a community they had visited as an adolescent on the move.

We also had examples from the Taï chimpanzees of similar though brief encounters between parties from neighbouring communities, including some sex between their members. This is not something we would have expected to find in chimpanzees, and these Taï communities have shown us another side to chimpanzee behaviour which comes close to that of bonobos.

Images of the brutal killing of chimpanzees by chimpanzees have been seared into our minds, alongside those of a non-violent bonobo. This has led to a belief in a greater polarisation of the two species than actually exists. While bonobos do show less of the more severe physical violence than chimpanzees, we have had plenty of evidence that severe violence does occur, and that females can be the perpetrators. This evidence for violence amongst the 'forgotten ape' appears to be in danger of being the most forgotten feature of the bonobo.

Adult bonobos have (heterosexual) sex all the time

We now know that heterosexual sex is not, in fact, especially frequent amongst sexually mature bonobos. Other than during the great agitation occurring in large groups due to discovering a large food source, or hearing vocalisations from a neighbouring community, or meeting up with other parties of the same community, sex amongst bonobos is a rare event.

For males, how much sex they get, even in these situations, depends on their rank, though opportunity can be greatly improved by having a 'wingmom'. We have discovered that male bonobos do compete for sexual access to females, and aggression and dominance pays off for them. There is less open aggression between bonobo males than chimpanzee males but this is due more to the fact that low-ranking male bonobos refrain from initiating sex, and thereby provoking higher status males, than the existence of an egalitarian sexual free-for-all.

Established female bonobos, who have more choice in the matter than do chimpanzee females, seem quite happy to endorse male sexual competition by mating more often with the high-ranking males. The young, low-status female bonobos have almost constant small sexual swellings and are the females most sexually active and least choosy; this is a strategy to avoid aggression and to gain access to food. These immigrant adolescent females will also not conceive for a number of years, so it is some time before they are going to be engaging in potentially fertile sex. Females with status are not as sexually active because they can more easily get what they want without the need to use sex to appease males. And these females can be particularly choosy about their sexual partners when conception is a possibility.

Established bonobo females in oestrus have a lower frequency of sex than their chimpanzee sisters, they can be quite choosy about their partners, and they generally show a lower proactive enthusiasm for sex. Yet these established mothers are the females with the highest status, and the ones that receive most respect from the males.

Kano (1992) states that female status does not come from continuous sexual receptivity, and he tells us that middle-aged to old females without swellings are more self-confident than the habitually swollen young females. Though males will be seeking mating opportunities it is not necessary for females to be in oestrus for males to seek association with them. So the idea that it is the extended sexuality of females that gives them their power by being constantly sexually attractive and sexually active does not fit with what actually occurs in bonoboland.

Even though there are normally more females with sexual swellings per male in a bonobo community than in a chimpanzee community, the males of both species are getting about the same amount of sex. Male and female bonobos traveling together in mixed parties are not necessarily engaging in much sexual activity, especially when those parties are small and therefore may not even include an oestrous female. Frances White at Lomako, for example, found that only 22% of foraging parties contained females with full sexual swellings (Stanford 1998).

Early captive studies featuring adolescent bonobo females created the image of constant sex amongst bonobos. The first studies at the artificial feeding sites in Wamba added to this image because large numbers of bonobos arriving at such a food source led to a lot of sex. When we get away from these early images of bonobos in zoos and at the feeding sites, we then arrive at a more realistic representation of this ape. Copulatory sex amongst bonobos turns out to be a rare event beyond certain specific circumstances, and bonobos are not, in fact, having sex all the time.

Bonobos are bisexual and pansexual, using sex like a hug or a handshake

Adult heterosexual sex is only half of the picture, and bonobos are also famous for their homosexual sex and for sex involving all ages. It is this casual pansexuality that the bonobo is particularly celebrated for, and it is this casual sex that is believed to have created an ape that makes love, not war. But we have found plenty of evidence from the field that bonobo life is neither so simple nor so sexy.

Sexual behaviour involving immatures has a lot in common with that of the chimpanzee, and the young males of both species are especially keen to get involved. Infant and juvenile bonobos have playful casual genital contacts amongst themselves but there is also some sexual contact between adults and immatures. When this involves an adult male with an infant it is playful, while those between mothers and their own infants can be about stress relief for the mother.

Juvenile males engage in a lot of sexual contact with adult females; more, in fact, than do the adult males. Genital contacts between adult males and juveniles are no longer as playful as they had been when the youngsters were infants, and now include more elements of dominance from those adult males.

When they reach adolescence, male bonobos are dominated by adult males and largely excluded from sex. Juvenile females and adolescent females still in their birth community are rarely involved in any sexual behaviour, while immigrant adolescent females are very sexually active. These differences in sexual activity due to age or sex of the individual would not exist if all this sexual behaviour was simply social sex or recreational sex; there is clearly more to it than that.

Some studies show mature males engaging in a lot of sociosexual activity with each other but other studies show little adult male homosexual interaction. In fact, the most common foraging parties of bonobos are small, and bonobos are not engaging in much sex of any description. The male-male mounting behaviours are not unlike those that occur in many species, and often involve a strong element of dominance, while rump-rubs and penis-rubbing are quite rare.

While all these male-male sociosexual behaviours are found to calm an agitated bonobo male, the behaviour is often forced by him. Rather than male homosexual behaviour being casual and friendly, it has turned out to be much more hostile, and often what we would call sexual assault. These males may be avoiding open 'war' but they are hardly 'making love'.

Female bonobos have also been seen to mate following aggression from the male, and even some gg-rubbings have been forced by the dominant female. Though gg-rubbings between females are not normally accompanied by open hostility or aggression, they are not about creating bonds of friendship either. These female gg-rubbings are mostly about status acknowledgement and the signalling of tolerance of the proximity of other females when feeding.

What is most apparent is that a lot of the sex amongst the mature bonobos is about relieving social tension rather than enjoying sex for its own sake. As Kano says, sex is induced by anxiety and seems unnecessary to bonobos when they are feeling relaxed. Sex between males or between females has not been found to be a bonding mechanism; it is a lubricant or grease that smooths the otherwise grating feeling that comes when proximity is forced between individuals who are not that fond of each other. Heterosexual genital contacts can also have the same function, and many of these genital contacts amongst bonobos stop short of how we would experience sex or expect sex to be.

We can say that a lot of bonobo sex is as casual and functional as our hugs and our handshakes, as long as we are especially talking about those hugs and those handshakes that occur in strained situations between people who don't really like each other much, if at all.

Bonobos engage in all kinds of sexual behaviour on a par with that of humans

When bonobo sexual behaviour is said to be on a par with that of humans, what is meant is that bonobos mate face-to-face, they use eye contact, they kiss, they engage in homosexuality, in threesomes and more, in masturbation, oral sex, and maybe even anal sex. The subject of sex between adults and youngsters is usually not included in this but is skirted around.

Those humans who are sexually attracted to children will find no defence or explanation amongst bonobos for that human attraction – not that the natural existence of a particular behaviour in another species is ever a reason to mark that behaviour as acceptable. There are many behaviours we abhor that can be found in nature; we only have to look at infanticide, or other painful or deadly aspects of sex and reproduction to see that natural does not equal desirable or good (Saxon 2012).

Whether something does or does not occur in nature cannot be used to prescribe or proscribe any behaviour in humans. But when we are trying to understand a particular human behaviour it can still be useful to look to other species to gain some insight as to why that behaviour might occur.

When it comes to bonobos, their pansexual behaviour is the way they handle social proximity and conflict. Because this sociosexual behaviour is important and widespread amongst adult bonobos, infants and other sexual immatures see adults engaging in such behaviours even if they are not directly involved to the same extent. Whether or not there is some innate predisposition to engage in such behaviour, bonobo youngsters do learn its use from watching the adults. The public nature of this bonobo sociosexual behaviour is one aspect of bonobo pansexuality that is definitely not on a par with that of humans.

As for all the other behaviours said to be enjoyed by bonobos, we have discovered that none are quite what we have been led to believe. Face-to-face sex is not something adult males do, oral sex has not been mentioned by field researchers, and anal penetration has been recorded once (Fruth and Hohmann 2006). As males often have erections during mounts of other males, and researchers have clearly stated time and again that penetration does not occur, we can rightly conclude that it is not something bonobos are intending.

Juvenile males may join in when an adult pair are engaged in sex but otherwise, threesomes or more are not for the bonobo. Eye contact does occur during face-to-face encounters, and females will sometimes look back at the male during doggy-style mounts (other primates such as macaques also do this). As for kissing, chimpanzees also kiss, even if it is not with the same enthusiasm as the juvenile bonobos at San Diego Zoo.

Most of these behaviours, along with masturbation and genital massage of another individual, have only come to be thought of as common bonobo behaviours because of captive studies, especially de Waal's San Diego study. Looking at the details of who was doing what to whom in that zoo population, we found that the juveniles were the main participants, and there was a particularly notable absence of their occurrence between the mature males and females.

Homosexual behaviour between males is neither common nor vastly different from that found in other primates. There are differences such as the rump-rubbing, occasional penis-rubbing, and the swapping of position during mounts, but dominance remains a clear aspect of this male homosexual behaviour, and shrieking victims are not very indicative of love-making.

Even the much more common homosexual behaviour between adult females is more about easing the discomfort of proximity than forming bonds. This is not about love-making either, nor is it about creating caring, sharing, egalitarian ties of friendship between females.

Adult bonobo sex is predominantly not play behaviour, engaged in just for fun or at times of chilled out relaxation. Though much of the sexual behaviour is not directly about reproduction it is about communication rather than recreation, and is used strategically and in particular circumstances. This behaviour is largely perfunctory and ritualistic, much like a hug or a handshake but signalling tolerance or calming an aggressor. These genital contact behaviours have their origin in sexual behaviour but their form and context have shifted, making then far less sexy than they first appear.

When we actually take a good look at bonobo sexual behaviour, is it really on a par with our own?

Or might we say, to paraphrase *Star Trek*'s Mr. Spock: "It's sex Jim, but not as we know it."

Chimpanzees are from Mars, bonobos are from Venus

Along with all these Venusian bonobo soundbites we also have the belief that chimpanzees are very different and very brutal. Chimpanzees can indeed be brutal, and chimpanzee males are clearly the dominant sex, but there is much variation across populations, and some populations, such as those in the Taï forest, show us that the distinction between the two species is not always so great.

Taï chimpanzees show a lot of association between the sexes and between females – females who can form strong friendships: supporting each other, sharing food with each other, and even adopting the weaned offspring of a friend if she dies. Many of the Taï females, like bonobo females, resume their sexual cycles within a year of giving birth (Boesch 2009).

Alpha males at Taï have also adopted orphaned youngsters, and have shown maternal behaviours towards them such as food-sharing, carrying, and sharing nests. Taï males have shown no inclination towards infanticide, neither in their own community nor in the many cases where they have engaged (including sexually) with isolated infant-carrying females from neighbouring communities.

Taï chimpanzees are preyed upon by leopards, and they will rush to support others who are being attacked. They show great concern for those who are injured, caring for their wounds by licking them clean. The killing of chimpanzees by chimpanzees is rare or absent across Taï communities, and this may have something to do with the eager support provided by community members when one of their number is under attack from a leopard. Whether the danger comes in the form of a predator or a party from a neighbouring community, help can be quickly on the scene.

We have also seen that sometimes during a stand-off between parties from neighbouring communities, females will cross over and mate with the males from the other community. This is something we did not expect to find amongst chimpanzees.

Dominant Taï females have been seen to support a son in his quest for alpha status (which also occurs in other chimpanzee populations). Taï females will also give social support to other males in their community, and they have more control over which males they mate with than we normally expect for chimpanzee females. They also play a role in ensuring that those males who have taken part in a successful hunt get their fair share of the meat, so they can have important and influential social roles.

Hunting has long distinguished chimpanzees from bonobos: while chimpanzees can be regular hunters of other primates, bonobos had only been seen to occasionally and opportunistically kill and eat small forest antelope, rodents, and perhaps a bird or two. Though killing other species for food does not make an animal more likely to kill members of its own species, the absence of intentional group hunting by bonobos has been associated with their less violent nature.

At LuiKotale there have now been three observed cases of successful hunting by bonobos where immature monkeys were caught, plus two failed attempts. Both sexes were active in the hunts, and one male and two females were the captors of the prey. The male did not share with others but one female shared with two adult females while the other shared with an adolescent male. Once again the gap between the two species draws closer.

Bonobos do not eat as much meat as chimpanzees but they do seem to relish what they can get. At Lomako, when an adult male bonobo captured a young duiker (forest antelope) he took it up in a tree, followed by the rest of the bonobo party. He quickly killed the duiker with a bite to the neck then he devoured the brain, and continued to eat half of the duiker without giving any to the begging females.

One of the females tried to pull off pieces of the meat, then she grabbed the back legs and there was a tug-of-war lasting several minutes. The female's juvenile daughter then jumped on the male from behind while screaming, and the male lost hold of the carcass. As he chased the young female her mother raced off with her prize which she then ate, giving only a little to her daughter.

At LuiKotale there has also been an observation of bonobo cannibalism. This was of an infant female, 2-3 years of age, and the cause of death is unknown. The mother was a lower-ranking female who was a more peripheral community member, and the body of the infant was initially taken from her and fed upon by a dominant female. The dominant female was soon joined in feeding by the infant's own mother. Six of the nine adult females who were present plus two of the three adult males consumed some of the infant over a period of about six hours, with possession of the carcass changing 14 times.

We cannot know whether anything untoward caused the death of the infant but this episode is another good reminder not to get carried away with an overly cute image of this ape.

There is a lot of variation across bonobos and chimpanzees, both as individuals and as communities, and rather than two clearly distinct species we find something that is much more of a continuum. To gain a better understanding of what goes on at the bonobo end of this continuum, we need to get past the over-hyped image of this ape: an image of an ape that is too busy having lots of incredibly pleasurable and chilled-out sex to bother with fighting. Looking past the sex we then get to see the bigger picture of bonoboland, and we discover the importance of some other aspects of their lives that are far less sexy but far more relevant when it comes to revealing the bonobo path to (relative) peace.

Weak male bonding

One aspect of bonobo life that has not, perhaps tellingly, become a popular soundbite is the weak bonding between males, yet this is a crucial part of life in bonoboland. It is the individualistic nature of male bonobos that stands out, yet this feature has been largely blotted out by the attention-grabbing image of pansexual bonding.

Looking past the sex, we can picture a community of bonobos foraging in small parties, the established mothers at the core of most of these groups. Immigrant females will seek out the senior females, while adult sons with mothers still living will often travel with their mother. Males in general will be coming and going between parties more frequently than the females, and if they are the dominant or the only mature male present then they will often be able to monopolise those females.

Some males, especially those of high-rank, will associate for various lengths of time with unrelated females and increase their chances of a fertile mating. Rather than random associations of equally attractive community members we have individuals who vary in how much they associate with others, expressing individual preferences and dislikes.

The bonobo foraging parties belong to a distinct community but it does not have the clear territorial boundary found in chimpanzees. Bonobo communities do have core areas but their territories overlap which means that the same food patches can be used by foraging parties from neighbouring communities. Chimpanzees, in contrast, have more distinct territorial boundaries patrolled by bands of males while females are in separate, small areas within that territory.

In chimpanzee communities, every so often a number of males will suddenly become quiet and set off in a line to patrol the boundary of their territory. This highly vigilant and silent line of males will travel for hours looking and listening for signs of their neighbours, and sometimes they will make incursions into neighbouring territory. Though actual contact is rare, this is deliberate action by male-only parties of chimpanzees along borders; there is the potential for conflict and, should a lone neighbour be encountered, the potential for a fatal attack.

Bonobo males do not carry out these border patrols. Instead, bonobo males are foraging alongside females in the overlapping areas of neighbouring communities, and this creates something very different from what we find in chimpanzees. Parties of males meeting parties of males from neighbouring communities are only going to be strangers to each other. But when we have this presence of female bonobos along community boundaries, we have the presence of bonobos who, unlike the males, have experienced life and other bonobos beyond the confines of the area in which they are now living.

This does not mean that female bonobos show no fear of the neighbours but it does mean that females will occasionally mingle with them. What matters for these bonobo parties is defending food patches, and this is best achieved by having as many members present as the food source allows because the larger party will win in any vocal contest over that food. Bonoboland is more about mixed-sex foraging parties defending food patches than bands of males defending, or trying to expand, a fixed territory.

Some adult males in these bonobo foraging parties are males traveling and foraging alongside their mother. Mother-son relationships are, Kano says, an obstacle to male bonding, but this should be a welcome obstacle when the absence of strong bonds between males is a major factor in the lower level of violence within communities as well as between them. Having a mother's support brings benefits to a male bonobo; a devoted mother is a much better ally than a fickle, unrelated male. We cannot know how established mothers came to be such dominant individuals in bonobo society but it could well be linked to the benefits of the mother-son relationship.

It is generally presumed that the status of females in bonobo society is due to their extended sexuality and sexual activity with males but this is unlikely to be so, considering how the females with the most extended sexuality and who are most sexually active with males are the bonobo females with the lowest status. The female bonobo's extended periods of sexual swellings may be linked more to relations between females than between the sexes, easing the tensions between established females and those who are recent immigrants, and allowing numbers of females to travel and forage together.

With females traveling together in parties, and with truly fertile sexual cycles difficult to clearly determine, male bonobos improve their mating and reproductive success by associating with their mothers, and by having good relations (relative to chimpanzees) with females in general. With the benefits to males of association with mothers and unrelated females, and without the motivation to form male alliances, the individualistic bonobo male has less reason to fight.

It is unlikely that fatal injuries have never occurred due to bonobo violence but bonobo males are much less violent than their chimpanzee brothers. This is not because they are getting more sex than their chimpanzee counterparts; it is because their road to sex is through good relations with females rather than alliances with males.

The bonobo is no longer the forgotten ape but we are in danger of falling for a bonobo imposter rather than seeing the real-life bonobo of the dark forests of the Congo. From a few early studies of small captive groups we were too quick to think we knew all there was to know about this other cousin of ours but she was, in fact, only gradually emerging from the shadows and revealing herself to us. As bonobo numbers dwindle and our research proceeds at its inevitably slow pace, perhaps we will never have the opportunity to fully know her but we do at least now know that there is far more to this rising star than the hyper-sexual, two-dimensional, media caricature.

I wanna be like you?

*H*aving bared all she can so far, our hippie starlet is now preparing to take her bow. We had eagerly anticipated tales of bonobo bliss: of wild sexual pleasure and a gentle, peaceful nature. Here was a sexy cousin, ready and waiting to rescue us from our unfortunate chimpanzee destiny. But our rising star has now revealed many shades of grey rather than a simple black-and-white opposition between these two cousins of ours. She has been brave to let us in on some of her darker secrets, and she has been a good sister in reminding us of the better side of chimpanzee nature.

As we have learned more about these two sister species we have discovered how much of an overlap there is, and the chimpanzee is no longer simply the evil twin. But the bonobo still resides mainly at the better end of the *Pan* continuum, where life is relatively more sexual and relatively more peaceful, and where females fare much better. The bonobo continues to provide us with an alternative to the chimpanzee story with its emphasis on male dominance, hunting, and warfare. And in the bonobo we still have a mother-centred cousin who uses sociosexual behaviour to deal with social tensions. How do we welcome her into our lives?

Most social media talk about the release of our inner bonobo is only (so far) talk about adults having lots of mutually pleasurable casual sex with other adults who turn them on, and how all this sexual pleasure leads to chilled-out dispositions, and feelings of goodwill to all others. The problem with this scenario is that it only exists in the wannabe's imagination, and is not what actually goes on in real-life bonoboland.

Bonobos use sexual or pseudosexual behaviour to alleviate many of the problems that arise between individuals in social groups but these are public conflicts, and immatures are either watching and learning or joining in to some degree. For bonobos there is nothing odd about engaging in genital contacts from birth. There is nothing especially odd about a stressed-out mother rubbing her infant between her legs, or an infant climbing aboard an adult male and rubbing genitals with him. Engaging the genitals is an ordinary part of bonobo life, and our young bonobo learns this from the adults who are expressing these sociosexual behaviours in front of the whole group. For our wannabe, on the other hand, this very public aspect of being a bonobo receives a sharp strike through.

Unfortunately for the wannabe, it also turns out that a lot of what we consider to be bonobo 'Kama Sutra' sex is actually the sexual behaviour of immature bonobos, especially the captive juvenile males. If immature sexual behaviour receives a strike through then many other bonobo behaviours are also gone with it, such as erotic kissing, oral sex, and threesomes. If the wannabe only wants to emulate the sexual behaviour of sexually mature bonobos (and it is, after all, the problem of conflict between grown-ups we are looking to solve) then they are left with something which is far less of a sexual smorgasbord; something rather more tame, in fact.

The wannabe is also left with the problem of how children are expected to learn bonobo-like sociosexual behaviour if they don't get to see it. When little Johnny is angry with little Jimmy we cannot really say "now, now, boys, time to mount each other" or "how about making up with a penis-rub?" It's not going to happen. When we seriously think about using sex and genital contacts as a much more bonobo-like form of social communication we are immediately faced with very obvious problems. In fact, the whole thing quickly becomes quite ridiculous.

The wannabe bonobo wants us to use sex like the bonobo – like a hug or a handshake – but to still hide these 'hugs' and 'handshakes' from the children. This rather undermines the idea that human sexual activity is really interchangeable with our hugs and handshakes. It would seem that children will still need to witness and learn the traditional human ways of dealing with conflict, such as thinking things through and using verbal communication – and even literal hugs and handshakes – while any quick genital rubs required throughout the day will have to be left to the adults in private. Perhaps special private booths would be set up for this purpose.

As children approach and enter puberty, and can finally be let in on that secret world of adult sex, perhaps they can then start to learn about the bonobo way. But our problems are not over yet. This is the time when the young female bonobo finally gets to unleash her swollen genitals on the world as she enters her new community of strangers. For these bonobo females (who will not be able to conceive for a number of years) it is about avoiding aggression from those strangers and getting close to the food. If the adolescent female wannabe is not going to be facing such social problems we don't know what to expect of her, though many adult wannabes will, no doubt, have plenty of helpful – and hopeful – suggestions.

Adolescent male bonobos, on the other hand, remain in their birth community but are now peripheral members of that community, and are subjected to a good deal of aggression. For sexually frustrated adolescent wannabe males there is not much solace to be had from the bonobo. After a very active sex life as a juvenile, the adolescent male bonobo hits something of a sexual brick wall. Now that he has become potentially fertile, the young male faces attacks from adult males, and a massive reduction in the willingness of adult females to engage in sex. It seems that hormone-fuelled adolescent males are everywhere condemned to face the misery of sexual frustration.

And so our discoveries continued. When we looked at how much sex sexually mature bonobos are getting we found that the frequency of heterosexual copulatory sex looked much the same as it is for chimpanzees. Both species are also promiscuous, mating with many different partners, so it is not promiscuity or the frequency of sex that leads to a more pacific male temperament. There are more bonobo females around with their sexy swollen rears but they – especially females with status – are not as keen to mate as their chimpanzee sisters. Male sexual competition is more subtle amongst bonobos rather than being absent, and female bonobos prefer the dominant males as mates. Bonobos are not having egalitarian, heterosexual sex all the time no matter how much the wannabe wishes it were so.

But surely, the fact that male bonobos are less violent than male chimpanzees has a lot to do with bonobo social interactions being more sexual (or, at least, more often involving the genitals), and this is the reason it works better for bonobos than the non-sexual ways of chimpanzee communication?

Maybe.

Calming an agitated or aggressive male by letting him copulate (even when that copulation is brief and non-ejaculatory) if you're female, or simply letting him rub his penis against you if you are a male (and sometimes if you are a female) might work better than a behaviour that does not reach and soothe a male so quickly. It is quite possible that some stimulation of the penis is the best way to switch off whatever the male might otherwise be feeling, thinking, or intending. But this does not mean that the partner in this genital stimulation is also experiencing pleasure; in some cases, such as the crouching, shrieking males, the pleasure is clearly not there.

Pacifying over-excited, aggressive males does not depend on this being a mutually pleasurable activity for females either. Young female bonobos who solicit sex in order to take food from a male are showing far less desire for the sex than for the food. Sometimes a female bonobo is charged at by a screaming male, his hair raised and teeth bared as he jumps on the female and bites at her. What does she do? The female, not surprisingly, screams but after retreating a short distance she will initiate sex to calm the male down. Is this what the female wannabe has had in mind?

So yes, sex to calm males down often works, but the wannabe would do well not to confuse this bonobo peacemaking behaviour with lovemaking. The best way to pacify a male may well be through his penis but wannabes need to address how they will incorporate this crucial bonobo 'social service' into their lives – unless this way of responding to outbursts of male aggression is another bonobo behaviour that is to receive a strike through.

What comes to the fore in bonobos is their homosexual activity. Bonobo sociosexual behaviour is bisexual behaviour, and there is no room for a wannabe to be just heterosexual or just homosexual. Social conflict does not occur only between two individuals who find each other sexually attractive. In fact, it mostly occurs between individuals who don't like each other much.

What will wannabe bonobos do when they feel like punching someone who does not turn them on?

Being a bonobo requires the ability to feel pleasure, or even just to give pleasure, by rubbing genitals with pretty much any other individual in your life. The situations where bonobos have the most sex are those where they are very stressed and agitated, and where a lot of aggressive behaviour is occurring. Avoiding an escalation of conflict is not about having sex with those you like; it's about having sex with those you don't like. Making love not war is not about sexual satiation leading to the absence of any negative feeling towards anyone else; it's about reacting to negative feelings towards another by rubbing genitals with them. If you feel annoyed or angry with someone, don't get mad, rub genitals.

We are, as we should be, searching for ways to end the human misery that comes from violence and war. Is the rubbing of genitals with all and sundry really going to be the answer? Are wannabes seriously thinking that people who are currently 'at war' in whatever way, whether with a neighbour, a work colleague, a stranger in the street, or someone from another culture or country, could rub their genitals together instead and all would be well? No. Wannabes are only thinking about getting all the sex they want with the people they find attractive.

This is not being a bonobo.

We started out thinking that it was all about the sex but the sociosexual behaviour of the bonobo is only one part of what makes bonobos different. The lower level of male competition and physical contact aggression compared to chimpanzees has also a lot to do with the relations between the sexes, especially those between mother and son.

Adult male bonobos will often travel with their mothers. For the lower-ranking males especially, this is how they can improve on their mating success. Mothers also seem more likely to seek higher status when they have sons approaching adulthood, and we have seen how violent these mothers can be. Not all males, of course, will have mothers who are still alive but when they do, this relationship is an important one. It may be that this close male-female relationship influences how males view females in general, and contributes towards the less dominant behaviour of males towards females.

It is the well-established mothers that have the status in bonoboland, and it is their sons who chiefly benefit. Daughters are gone once they reach puberty, and immigrant females have low status. It is not easy to see how this translates into the expectations of wannabes. What happens when daughters are still around? How close do male wannabes really want to be with their mothers? And how much do female wannabes want to focus on their sons, perhaps sacrificing relationships with their daughters in order to build that close mother-son connection, and so weaken those potentially dangerous bonds between males?

Our wannabe bonobos often either don't have children or don't want children. This, of course, is not the way of the bonobo and changes the whole dynamic of the male-female relationship. Female wannabe bonobos who are not yet mothers or have chosen not to be mothers have more in common with the low-status adolescent female bonobo than the high-status mothers of multiple offspring.

For the bonobo female who has produced offspring, especially when she has a son, those offspring are what motivates her behaviour. She will aggressively protect her young, and she may aggressively, even violently, act to improve the status of her sons. The sons are surely a strong motivation for her power-seeking behaviours. And she has to balance her interests as a female with those of her male offspring.

Though female wannabes may not have sons, male wannabes will usually still have mothers around. How much does the male wannabe want his mother to be his main ally and companion; how much does he want an ever-present wingmom? His inner bonobo should be pushing him in that direction and away from relationships with other males. If we also have to strike through this aspect of bonobo behaviour we are losing something of the greatest importance when it comes to our attempts to emulate the bonobo path to peace.

Weak male bonding is the most essential factor in the lower levels of male domination and male violence in bonoboland. Male teamwork is a chimpanzee trait. It is the individualistic nature of male bonobos that has, as de Waal says, paved the way for a collective female power takeover (de Waal 1997). If our male wannabes play or follow team sports they are expressing their inner chimpanzee, not their inner bonobo. To strike through this aspect of the bonobo is the final death knell for the bonobo way.

When it comes down to it, do wannabe bonobos really want to release their inner bonobo? We discovered that most of that 'Kama Sutra' smorgasbord of bonobo sexual behaviour is not adult sexual behaviour. Then we have had to strike through many aspects of bonobo behaviour such as sex between immatures and matures, forced homosexual male sex, weak bonds between males, and sex with those we don't like. There are other aspects we are not that happy about either, such as adolescent female sex used in exchange for food or to avoid aggression, inequality of access to sex, female aggression and violence, and female status leading to lower sexual activity. There is hardly anything of the bonobo left.

We are an ape, a naked ape, but we are not a naked bonobo any more than we are a naked chimpanzee or a naked gorilla. We have had our own, unique evolutionary journey, and while we can seek clues from other apes, and from other species beyond the apes, we will remain uniquely human.

Could we manipulate human behaviour to produce less violent males? Male bonding and hunting have been, whether we like it or not, a significant part of our evolution. If we look to the bonobo for help, preventing male bonding is the most crucial piece of their relatively peaceful behaviour they have to offer, and there's the rub — one rub that really should be grabbing our attention. And no amount of pseudo-bonobo gg-rubbing or rump-rubbing will make this particular obstacle to peace go away. The naked ape is going to have to find another way.

The major chimpanzee and bonobo research sites

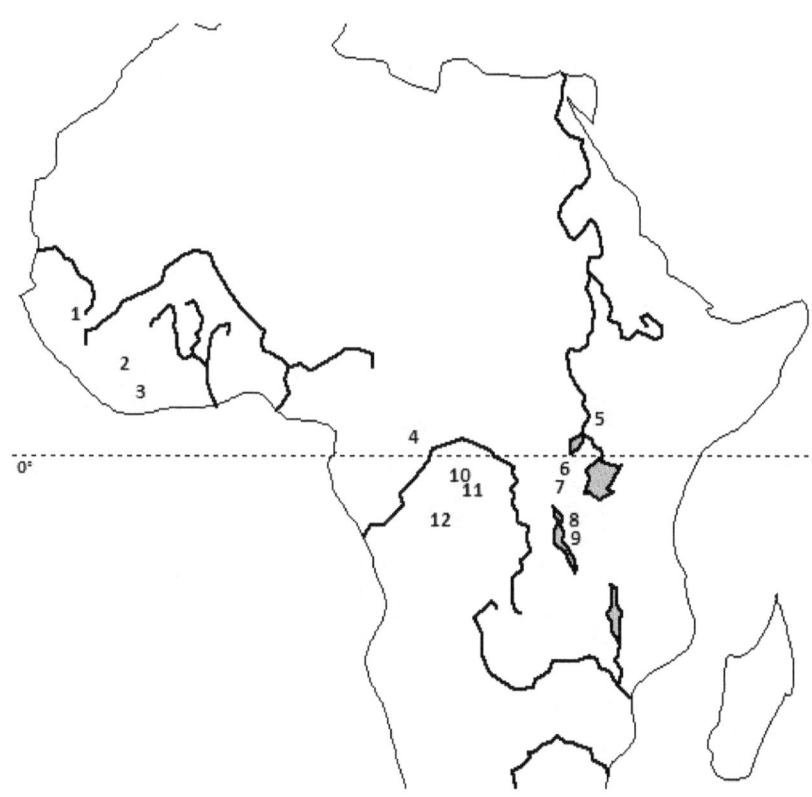

1-9 Chimpanzees:

1. Fongoli	6. Kibale
2. Bossou	7. Kalinzu
3. Taï	8. Gombe
4. Goualougo	9. Mahale
5. Budongo	

10-12 Bonobos:

10. Lomako
11. Wamba
12. LuiKotale

References

Badrian, A., and Badrian, N. (1984). Social Organization of *Pan paniscus* in the Lomako Forest, Zaire. In R. L. Susman (Ed.), *The Pygmy Chimpanzee: Evolutionary Biology and Behavior.* New York: Plenum Press.

Blount, B. G. (1990) Issues in Bonobo (*pan paniscus*) Sexual Behavior. *American Anthropologist,* New Series, 92(3): 702-714.

Boesch, C. (2009). *The Real Chimpanzee: Sex Strategies in the Forest.* Cambridge: Cambridge University Press.

de Lathouwers, M., and van Elsacker, L. (2006). Comparing infant and juvenile behavior in bonobos (*Pan paniscus*) and chimpanzees (*Pan troglodytes*): a preliminary study. *Primates*, 47(4): 287-93.

de Waal, F. B. M. (1989). *Peacemaking among Primates.* Cambridge, MA: Harvard University Press.

de Waal, F. B. M. (1995). Sex as an Alternative to Aggression in the Bonobo. In P. R. Abramson, and S. D. Pinkerton (Eds.), *Sexual Nature Sexual Culture.* Chicago: university of Chicago Press.

de Waal F. B. M. (2001). Apes from Venus: Bonobos and Human Social Evolution. In F. M. de Waal (Ed.), *Tree of Origin: What Primate Behavior Can Tell Us about Human Social Evolution.* Cambridge, MA: Harvard University Press.

de Waal F. B. M. (2005). *Our Inner Ape: The Best and Worst of Human Nature.* London: Granta books.

de Waal, F. B. M., and Lanting, F. (1997). *Bonobo: The Forgotten Ape.* Berkeley, CA: University of California Press.

Dixson, A. F. (2012). *Primate Sexuality: Comparative Studies of the Prosimians, Monkeys, Apes, and Human Beings.* 2nd Edition. Oxford: Oxford University Press.

Fruth, B., and Hohmann, G. (2006). Social grease for females? Same-sex genital contacts in wild bonobos. In V. Sommer and P. L. Vasey (Eds.), *Homosexual Behaviour in Animals: An Evolutionary Perspective.* Cambridge: Cambridge University Press.

Furuichi, T. (1987). Sexual swelling, receptivity, and grouping of wild pygmy chimpanzee females at Wamba, Zaïre. *Primates,* 28(3): 309-318.

Furuichi, T. (1997). Agonistic Interactions and Matrifocal Dominance Rank of Wild Bonobos *(Pan paniscus)* at Wamba. *International Journal of Primatology,* 18(6): 855-875

Furuichi, T., and Hashimoto, C. (2002). Why female bonobos have a lower copulation rate during estrus than chimpanzees. In C. Boesch, G. Hohmann, and L. F. Marchant (Eds.), *Behavioural Diversity in Chimpanzees and Bonobos.* Cambridge: Cambridge University Press.

Furuichi, T., and Hashimoto, C. (2004). Sex differences in copulation attempts in wild bonobos at Wamba. *Primates,* 45(1): 59-62.

Gerloff, U., Hartung, B., Fruth B., Hohmann, G., and Tautz, D. (1999). Intracommunity relationships, dispersal pattern and paternity success in a wild living community of bonobos *(Pan paniscus)* determined from DNA analysis of faecal samples. *Proc. R. Soc. B.,* 266(1424): 1189-1195.

Goodall, J. (1968). The Behaviour of Free-living Chimpanzees in the Gombe Stream Reserve. *Animal Behaviour Monographs,* 1:161-311.

Harris, E. E. (2015). *Ancestors in our Genome: The New Science of Human Evolution.* New York: Oxford University Press.

Hashimoto, C. (1997). Context and Development of Sexual Behavior of Wild Bonobos (*Pan paniscus*) at Wamba, Zaire. *International Journal of Primatology*, 18(1): 1-21.

Hashimoto, C., and Furuichi, T. (1994). Social Role and Development of Noncopulatory Sexual Behavior of Wild Bonobos. In R. W. Wrangham *et al.* (Eds.), *Chimpanzee Cultures*. Cambridge, MA: Harvard University Press.

Hohmann, G., and Fruth, B. (2000). Use and function of genital contacts among female bonobos. *Animal Behaviour*, 60(1): 107-120.

Hohmann, G., and Fruth, B. (2002). Dynamics in social organization of bonobos (*Pan paniscus*).) In C. Boesch, G. Hohmann, and L. F. Marchant (Eds.), *Behavioural Diversity in Chimpanzees and Bonobos.* Cambridge: Cambridge University Press.

Hohmann, G., and Fruth, B. (2003). Intra- and inter-sexual aggression by bonobos in the context of mating. *Behaviour*, 140(11/12): 1389–1413.

Hohmann, G., Gerloff, U., Tautz, D., and Fruth, B. (1999). Social bonds and genetic ties: kinship, association and affiliation in a community of bonobos (*Pan paniscus*). *Behaviour,* 136(9): 1219-1235.

Idani, G. (1990). Relations between unit-groups of bonobos at Wamba, Zaire: encounters and temporary fusions. *African Study Monographs,* 11(3): 153-186.

Jahme, C. (2001). *Beauty and the Beasts: Woman, Ape and Evolution.* London: Virago.

Kano, T. (1990). The Bonobo's Peaceable Kingdom. *Natural History,* 11: 62-71.

Kano, T. (1992). *The Last Ape.* Stanford, CA: Stanford University Press.

Kano, T. (1996). Male rank order and copulation rate in a unit-group of bonobos at Wamba, Zaire. In W. C. McGrew *et al.*

(Eds.), *Great Ape Societies.* Cambridge: Cambridge University Press.

Kano, T., and Mulavwa, M. (1984) Feeding Ecology of the Pygmy Chimpanzees (*Pan paniscus*) of Wamba. In R. L. Susman (Ed.), *The Pygmy Chimpanzee: Evolutionary Biology and Behavior.* New York: Plenum Press.

Kitamura, K. (1989) Genito-genital contacts in the pygmy chimpanzee (*pan paniscus*). *African Study Monographs,* 10(2): 49-67.

Legrain, L., Stevens, J., Alegria Iscoa, J., and Destrebecqz, A., (2011). A case study of conflict management in bonobos: how does a bonobo (Pan paniscus) mother manage conflicts between her sons and her female coalition partner? *Folia Primatol.,* 82(4-5): 236-43.

Morris, D. (1967). *The Naked Ape: A Zoologist's Study of the Human Animal.* London: Jonathan Cape.

Myers Thompson, J. A. (2002). Bonobos of the Lukuru Wildlife Research Project. In C. Boesch, G. Hohmann, and L. F. Marchant (Eds.), *Behavioural Diversity in Chimpanzees and Bonobos.* Cambridge: Cambridge University Press.

Nadler, R. D. (1986). Sex-related behavior of immature wild mountain gorillas. *Developmental Psychobiology,* 19(2): 125-137.

Parish, A. R. (1996). Female relationships in bonobos (*Pan paniscus*): evidence for bonding, cooperation, and female dominance in a male-philopatric species. *Human Nature,* 7: 61-96.

Parker, I. (2007). Swingers: bonobos are celebrated as peace-loving, matriarchal, and sexually liberated. Are they? *The New Yorker,* Jul 30: 48-61.

Prüfer, K., Munch, K., Hellmann, I., *et al.* (2012). The bonobo genome compared with the chimpanzee and human genomes. *Nature,* 486(7404): 527-31.

Ryu, H., Hill, D. A. and Furuichi, T. (2014). Prolonged maximal sexual swelling in wild bonobos facilitates affiliative interactions between females. *Behaviour,* DOI:10.1163/1568539X-00003212.

Saxon, L. (2012). *Sex at Dusk: Lifting the Shiny Wrapping from Sex at Dawn.* Printed by Createspace.

Stanford, C. B. (1998). The Social Behavior of Chimpanzees and Bonobos: Empirical Evidence and Shifting Assumptions. *Current Anthropology,* 39(4): 399-420.

Stevens, J. M., Vervaecke, H., de Vries, H., and van Elsacker, L. (2006). Social structures in Pan paniscus: testing the female bonding hypothesis. *Primates,* 47(3): 210-7.

Stevens, J. M., Vervaecke, H., and van Elsacker, L. (2008). The Bonobo's Adaptive Potential: Social Relations Under Captive Conditions. In T. Furuichi and Jo Thompson (Eds.), *The Bonobos: Behavior, Ecology and Conservation.* New York: Springer.

Surbeck, M., Deschner, T., Schubert, G., Weltring, A., and Hohmann, G. (2012). Mate competition, testosterone and intersexual relationships in bonobos, *Pan paniscus. Animal behaviour,* 83(3): 659-669.

Surbeck, M., and Hohmann, G. (2013). Intersexual dominance relationships and the influence of leverage on the outcome of conflicts in wild bonobos (*Pan paniscus*). *Behav. Ecol. Sociobiol.* 67(11): 1767-1780.

Surbeck, M., Mundry, R., and Hohmann, G. (2010). Mothers matter! Maternal support, dominance status and mating success in male bonobos (*Pan paniscus*). *Proc. R. Soc. B.,* 278: 590-598.

Takahata, Y., Ihobe, H., and Idani, G. (1996). Comparing copulations of chimpanzees and bonobos: do females exhibit proceptivity or receptivity? In W. McGrew, L. F. Marchant, and T. Nishida (Eds.), *Great Ape Societies.* Cambridge: Cambridge University Press.

Thompson-Handler, N., Malenky, R. K., and Badrian, N. (1984). Sexual Behavior of Pan paniscus under Natural Conditions in the Lomako Forest, Equateur, Zaire. In R. L. Susman (Ed.), *The Pygmy Chimpanzee: Evolutionary Biology and Behavior.* New York: Plenum Press.

Tutin, C. E. G., and McGinnis, P. R. (1981). Chimpanzee Reproduction in the Wild. In C. E. Graham (Ed.), *Reproductive Biology of the Great Apes: Comparative and Biomedical Perspectives.* New York: Academic Press.

Vervaecke, H., Stevens, J., and van Elsacker, L. (2003). Interfering with Others: Female-Female Reproductive Competition in *Pan paniscus.* In Clara B. Jones (Ed.), *Sexual Selection and Reproductive Competition in Primates: New Perspectives and Directions.* American Society of Primatologists, Norman, Oklahoma.

White, F. J. (1992). Eros of the Apes. *BBC Wildlife Magazine* 10(8): 38-47.

White, F. J., and Lanjouw, A. (1992). Feeding competition in Lomako bonobos: Variation in social cohesion. In T. Nishida *et al.* (Eds.), *Topics in Primatology Vol. I. Human Origins.* Tokyo: University of Tokyo Press.

Woods, V. and Hare, B. (2010). Bonobo but not chimpanzee infants use socio-sexual contact with peers. *Primates,* 52(2): 111-6.

Wrangham, R. W., and Glowacki, L. (2012). Intergroup aggression in chimpanzees and war in nomadic hunter-gatherers: evaluating the chimpanzee model. *Human Nature,* 23(1): 5-29.

Wrangham, R. W., and Peterson, D. (1997). *Demonic males: Apes and the Origins of Human Violence.* London: Bloomsbury.